# Lecture Notes in Mathematics

Edited by A. Dold and B. Eckmann

927

Yuval Z. Flicker

# The Trace Formula and Base Change for GL(3)

Springer-Verlag
Berlin Heidelberg New York 1982

**Author**
Yuval Z. Flicker
Department of Mathematics, Princeton University
Fine Hall – Box 37, Princeton NJ 08544, USA

AMS Subject Classifications (1980): 10D40, 12A85, 22E50, 22E55

ISBN 3-540-11500-5 Springer-Verlag Berlin Heidelberg New York
ISBN 0-387-11500-5 Springer-Verlag New York Heidelberg Berlin


Printed in Germany

Printing and binding: Beltz Offsetdruck, Hemsbach/Bergstr.
2141/3140-543210

Math

Sep.

---

Table of Contents

## §0. INTRODUCTION

The experience gained in studies of automorphic forms on GL(2) and related groups (such as its inner forms, SL(2) and n-fold covering groups of GL(2) ) suggests that any extension of the theory to other algebraic groups will be accompanied, or rather depend on, a further insight into the trace formula. One of our aims in this paper is to present a new way of writing the non-elliptic terms ("cusps") of the trace formula of Jacquet-Langlands [8], Langlands [12] and Arthur [1,2,3]. The other is to apply it in the case of GL(3) to the study of the base change problem, generalizing the results of Langlands [12] for GL(2). From the complete expression for the formula (singular terms included) which is established here, using [3] we also obtain the trace equality which affords relating automorphic representations of GL(3) and its inner forms.

The problem of base change is that of relating, or lifting, automorphic forms $\pi$ on a reductive group G over a number field F and those $\pi^E$ on G over a finite extension E of F . The theory of Langlands [12], Saito and Shintani [15] describes such liftings only when the extension E/F is cyclic and G is the group GL(2) . These relations can be predicted by Langlands' principle of functoriality for automorphic forms [12]. In practice they first occurred in the work of Doi and Naganuma who applied the Hecke theory to prove the existence of a lifting $\pi^E$ when F = Q and E is real quadratic. Jacquet used the representations of [8] to remove the restrictions on F, E, and $\pi$ , but the method of L-functions could not characterize the $\pi^E$ which were liftings, and its scope was restricted to quadratic field extensions.

Following Saito who introduced the initial idea in the context of modular forms, Shintani [15] suggested using the twisted trace formula in the context of automorphic representations to remove the restrictions inseparable from the techniques of modular forms. The techniques required to carry out that project were developed by Langlands [12] on the basis of [8], exploiting the Bruhat-Tits buildings to establish the required traces identity and using the orthogonality relations to derive the local and global base change theory for $GL(2)$. The complete work established the existence and characterized the liftings for any cyclic extension $E/F$ of prime degree $\ell$, and was then applied to the study of the Artin L-function for some non-induced representations of the Weil group. The main application of our study of the trace formula here is to generalize the theory to $GL(3)$, thus proving that when $E/F$ is a cyclic extension of prime degree of global or local fields of characteristic 0 then each automorphic (locally: admissible) representation of $GL(3,F)$ lifts, or corresponds, to such representation of $GL(3,E)$. We also characterize the image of the correspondence and describe its fibers.

A first attempt at generalization was made by Arthur [1] who obtained an expression for the trace formula for arbitrary reductive group as a sum of terms indexed by regular and singular equivalence classes of elements in $G(F)$. Explicit expressions were given in [1] for the regular classes. For any applications we must have explicit expressions for all terms. In our case of $GL(3)$ such expressions are found in Lemmas 2.3 and 2.4 (and proved in Sections 4 and 5 of Chapter 2).

To explain some of the complications which arise in the higher rank case we recall that the main step in [12] is a comparison of the trace formula for $G(F)$ and the twisted trace formula for $G(E)$. When all easy cancellations in the difference of the formulae were observed the Poisson summation formula was applied to the remainder which was expressed in terms of invariant distributions. The last point was taken up by Arthur [3] who inductively expressed the terms in

[1] as invariant distributions. However the Poisson summation formula was not applied in [3] and the question of the use of [3] to our applications was left untouched. In our opinion appealing to the summation formula at the right moment is fundamental for applications. Our purpose here is to suggest a way of invoking it and to carry out this suggestion in our case of GL(3).

Various difficulties arise when attempting to apply the summation formula here. One of them is that the smooth function to which this formula is usually applied does not exist. The natural candidate is the term in the trace formula indexed by a regular non-elliptic conjugacy class, viewed as a function on the split component of the torus in question. In the case of the split classes not only is this function non-smooth but it is also singular near the singular set. It is given by an orbital integral weighted by a factor whose degree is the split rank of $G$. For GL(2) the rank is $1$, the weight factor is linear and so the difference between this function and its analogue from the twisted trace formula is smooth (at all finite places; see [12]). When the rank exceeds $1$ this observation cannot be made and at first glance it appears that the bad behavior of the function must indeed prevent us from applying the summation formula.

To overcome this difficulty using the product formula on a number field we "correct" the global weighted orbital integral without changing its value on the F-rational points under consideration in such a way that the resulting function is no longer singular near the singular set. The fact that a single global correction suffices to make each of the local components regular is remarkable. But this alone would not suffice since the resulting function although regular is non-smooth. Our only hope is that our function will nevertheless be amenable to the application of the Poisson summation formula, which is valid for any compactly supported continuous function whose Fourier transform is integrable.

Thus we study the asymptotic behavior of the corrected weigted orbital integrals at the singular set. An additional lemma (2.8) shows that the summation formula can indeed be applied to such non-smooth functions with the given asymptotic behavior. Even all this would not have been enough if the limits at the singular set were missing. Fortunately they do appear naturally in the terms of the trace formula indexed by the singular classes. Even in the case of GL(2) the introduction of the correction and the application of the summation formula to the resulting non-smooth function significantly simplifies the work (cf. [12], Chapter 9). After this work was written the same "correction" was applied in some other cases (e.g. [5]).

A substantially different discussion is carried out for the quadratic terms in the trace formula. The summation formula is now applied on the split component of the associated quadratic tori. Passing to a local problem we resort to the results of [12] on spherical functions for GL(2) and apply a Plancherel formula for spherical functions to obtain a suitable expression. Although they suffice for our application the methods employed here are perhaps only provisional. This may also be the case with [3] where contributions from the continuous spectrum were incorporated with those indexed by equivalence classes in G(F) to express the trace formula in terms of invariant distributions. In a more mature theory a dual approach to [3] will probably be followed, namely that first the Poisson formula will be applied (with respect to the split component) to the terms indexed by non-elliptic classes in G(F) , corrected as suggested here, and only then will the Selberg trace formula be inductively applied to lower rank subgroups to combine the result with the terms from the continuous spectrum to form the invariant districutions. With a little more attention to the asymptotic behavior of the (corrected) quadratic terms near 1 (which we do not need here) such a project can probably be carried out now for GL(3) . For other groups the singular terms and the asymptotic behavior of the regular terms near the singular set will have to be studied, or at least their properties temporarily assumed, before this can be done.

The trace formula for GL(3) and its regular terms are recalled in Chapter 2 (from [1,3]), where we calculate the singular terms, introduce the corrected weighted orbital integrals, study their asymptotic behavior (at the finite places) and relate the limits to the singular terms of the trace formula. In Lemma 2.8, we prove that the summation formula applies to such functions. The analogues of these results for the twisted trace formula are dealth with in Chapter 3. The twisted analogue of [3] (cf. 3.2.4) is only sketched. This is a subject for a separate paper which we believe should not be written out now but only together with the alternative approach for the trace formula suggested above. In any event all of the results that we need can be obtained on tracing the effect of the twisting by $\sigma$ through [1,2,3] (as we did); this is in contrast to [5] where the outer twisting of that paper leads to a substantially different outerly twisted trace formula.

In Chapter 4 we consider the difference of the contributions to the trace formula from the continuous spectra. This difference will describe in Chapter 6 the continuous series representations of G(E) obtained from cuspical representations of G(F) by the base change correspondence. Also in Chapter 4 we express a discrete sum of traces of representations as a discrete sum of values of the Satake transform of some spherical component of a global function. Note that a full expression for the trace formula is given in the union of Lemma 2.7 and Chapter 4 (twisted formula is in Lemma 3.11 and Chapter 4). It is in Chapter 5 where we apply the Poisson formula to obtain from the remains of the difference of trace formulae a continuous sum, and special attention is given to the quadratic terms. The required traces relation follows by contradicting an equality between discrete (Lemma 4.3) and continuous (Lemma 5.4) measures (unless both are 0) using spherical functions. In contrast to [12], where such an argument was first employed, our treatment of the quadratic terms forces us not only to approximate the Satake transform (as in [12]) but also its derivative.

In addition to the global results from the theory of the trace formula we use in the final chapter, where the local global base change theory is established, the local results of Chapter 1. They consist of results on characters of representations, notably some orthogonality relations for characters of square-integrable representations, and results concerning orbital integrals. There are no difficulties at a place $v$ of $F$ which splits in $E$ (although we could not prove directly Lemma 1.15 for an arbitrary $\ell$-tuple of spherical functions). At other $v$ we have to match functions $\phi_v$ and $f_v$ with the same orbital integrals on $G(E_v)$ and $G(F_v)$. For the proof of Proposition 5.5 it is important to know that when $v$ is unramified in $E$, and $\phi_v$, $f_v$ are matched ($\phi_v \longrightarrow f_v$), and one of $\phi_v$, $f_v$ is spherical, then the other can also be assumed to be spherical. This was proved by Kottwitz [11] using the Bruhat-Tits building for $SL(3)$, generalizing [12], §5.

For the study of $\phi_v \longrightarrow f_v$ when $\phi_v$ and $f_v$ are not necessarily spherical we use the classification of orbital integrals for $GL(3,F_v)$. Generalizing the corresponding result of [12] for $GL(2)$ this was deduced by Flath [4] from a (known) case of Howe's conjecture when $F_v$ is non-archimedean. In this context we note that the main step in the proof of the global result asserted in [4] (which is a certain equality of trace formulae for GL(3) and the multiplicative group of a division algebra of dimension $3^2$ over F ) was stated without proof in [4]. This is established in Corollary 2.9 here, using also the splitting property of [3], §11 (in Lemma 2.6 below). The deduction of the correspondence in the case of division algebras is much easier and the reader of Chapter 6 below will benefit from reconstructing it, using the ("generalized") linear independence of characters from [8] (cf. end of proof of Lemma 6.3).

Our base change theory is incomplete in the case where there is an archimedean place of F which ramifies in E (hence $\ell = 2$ , $F_v = \mathbb{R}$ , $E_v = \mathbb{C}$ for the place v ), since the classification of orbital integrals of compactly supported functions on $GL(3,\mathbb{R})$ is only conjectured in Lemma 1.2. As noted above we are not concerned here with writing out the details of the proof of the twisted analogues of [1,2,3]. We merely traced the effect of the twisting by $\sigma$ through these papers, recording the results which we need (3.2.4; cf. [5] where the outerly twisted trace formula is substantially different from the trace formula). Finally note that the details of the exercises of 2.7.1/2 were given only in what we feel is adequately inexcessive fashion (including a reference to [12], §9 , for the archimedean places).

It remains for me to thank J. Arthur, H. Jacquet and R. P. Langlands for invaluable advice while preparing this work.

## §1. LOCAL THEORY

### 1.1.1 Notations

Let $\ell$ be a prime, $F$ a number field, $E$ a cyclic extension of $F$ of degree $\ell$, $G$ the galois group of $E$ over $F$. For each place $v$ of $F$ denote by $F_v$ the completion of $F$ at $v$ and put $E_v = E \otimes_F F_v$. Since $E/F$ is cyclic $v$ may either split completely in $E$, in which case $E_v$ is isomorphic to a direct sum of $\ell$ copies of $F_v$ and $G$ acts by permuting the components, or $v$ stays prime or ramifies in $E$, and then $E_v$ is a field and the galois group of $E_v/F_v$ is isomorphic to $G$. If $\ell$ is odd the infinite places of $F$ always split completely in $E$.

When dealing with a fixed local field we shall drop the index $v$. For example if $F$ is local we denote by $\alpha$ the unramified character $\alpha(x) = |x|$ on $F$ (the valuation is normalized so that the product formula is valid). If $F$ is non-archimedean $\tilde{\omega}$ denotes the local uniformizing parameter in the ring $\mathcal{O}$ of integers in $F$. We write $F^\times$ for the multiplicative group of $F$. By $\mathbb{A}$, $\mathbb{A}^\times$, $\mathbb{A}_E$, $\mathbb{A}_E^\times$ we denote the adeles of $F$, ideles of $F$, adeles of $E$, ideles of $E$.

Put $G = GL(3)$. It is convenient to choose the notations $P_0, P_1, P_2$ for the standard parabolic subgroups of type (1,1,1), (2,1), (1,2), that is

$$P_0 = \left\{\begin{pmatrix} * & * & * \\ 0 & * & * \\ 0 & 0 & * \end{pmatrix}\right\}, \quad P_1 = \left\{\begin{pmatrix} * & * & * \\ * & * & * \\ 0 & 0 & * \end{pmatrix}\right\}, \quad P_2 = \left\{\begin{pmatrix} * & * & * \\ 0 & * & * \\ 0 & * & * \end{pmatrix}\right\}.$$

Let $M_i$, $A_i$, $N_i$ be the Levi subgroup, split component of $M_i$, and the unipotent radical, of $P_i$ $(0 \leq i \leq 2)$. For example

$$M_1 = \left\{\begin{pmatrix} * & * & 0 \\ * & * & 0 \\ 0 & 0 & * \end{pmatrix}\right\}, \quad A_1 = \left\{\begin{pmatrix} a & & 0 \\ & a & \\ 0 & & b \end{pmatrix}\right\}, \quad N_1 = \left\{\begin{pmatrix} 1 & 0 & * \\ & 1 & * \\ & & 1 \end{pmatrix}\right\}.$$

The centre of $G$ is denoted by $Z$. If $R$ is a ring and $H$ is an algebraic group defined over $R$ we denote by $H(R)$ or by $H_R$ the R-rational points on $H$. For each $v$ we choose a maximal compact subgroup $K_v$ of $G(F_v)$, namely $G(\mathcal{O}_v)$ if $v$ is finite, $O(3,\mathbb{R})$ if $F_v$ is the field $\mathbb{R}$ of real numbers, and $U(3,\mathbb{R})$ if $F_v$ is the field $\mathbb{C}$ of complex numbers.

1.1.2 The norm map

Let $F$ be either a local or a global field, $E$ a cyclic galois extension of prime degree $\ell$, and $\sigma$ a fixed generator of the galois group $\mathcal{G}$ of the extension $E/F$. The elements $\gamma$, $\delta$ of $G(E)$ are said to be $\sigma$-conjugate if $\delta = g^{-\sigma}\gamma g$ for some $g$ in $G(E)$. The relation $g^{-1}N\gamma g = N(g^{-\sigma}\gamma g)$ shows that

$$\gamma \longmapsto N\gamma = \gamma^{\sigma^{\ell-1}} \cdots \gamma^{\sigma}\gamma \quad (\gamma \text{ in } G(E))$$

is a map from the set of $\sigma$-conjugacy classes in $G(E)$ to the set of conjugacy classes in $G(E)$ and that any conjugate of $N\gamma$ is obtained by the map. The identity $(N\gamma)^{\sigma} = \gamma(N\gamma)\gamma^{-1}$ shows that the set of eigenvalues of $N\gamma$ is invariant under $\mathcal{G}$ and hence $N\gamma$ is conjugate over $G(E)$ to an element of $G(F)$. The map is one-to-one since $\gamma$ and $\delta$ are $\sigma$-conjugate whenever $N\gamma$ and $N\delta$ are conjugate (see below). Thus we obtain an injection from the set of $\sigma$-conjugacy classes in $G(E)$ into the set of conjugacy classes in $G(F)$.

We have to determine the image of the map. Here the observation that $\gamma$ (in $G(E)$) lies in the centralizer $G_h(E)$ of $h = N\gamma$ if $h$ lies in $G(F)$, is fundamental. We shall list a set of representatives $h$ for the conjugacy classes in $G(F)$ and describe those for which $h$ is of the form $N\gamma$ for some $\gamma$ in $G(E)$.

(1) $h = \alpha\begin{pmatrix}1 & a & b\\ & 1 & c\\ 0 & & 1\end{pmatrix}$ with $ac \neq 0$ or $a = c = 0$, $b \neq 0$. Then $h$ is of the form $N\gamma$ if and only if $\alpha$ lies in $NE^\times$. Indeed, if $h = N\gamma$ then $\gamma$ lies in $G_h(E)$ and has the form $\gamma = \beta\begin{pmatrix}1 & p & r\\ & s & q\\ & & 1\end{pmatrix}$. Hence $N\gamma = N\beta\begin{pmatrix}1 & * & *\\ & Ns & *\\ & & 1\end{pmatrix}$ and $\alpha = N\beta$. In the opposite direction we note that if $\alpha = N\beta$ then $h = N\left(\beta\begin{pmatrix}1 & & b/\ell\\ & 1 & \\ & & 1\end{pmatrix}\right)$ if $a = c = 0$, and for the case when $ac \neq 0$ we note that

$$N\left(\beta\begin{pmatrix}1 & p & r\\ & 1 & q\\ & & 1\end{pmatrix}\right) = N\beta\cdot\begin{pmatrix}1 & \mathrm{tr}p & \mathrm{tr}r+f(p,q)\\ & 1 & \mathrm{tr}q\\ & & 1\end{pmatrix}$$

where $f(p,q)$ is a polynomial in the conjugates of $p$ and $q$.

(2) $h$ in $Z(F) = F^\times$. If $\ell = 3$ then $h = N\gamma$ always has a solution. If $\ell \neq 3$ then $h = N\gamma$ if and only if $h$ is in $NE^\times$. The first statement follows from $N\begin{pmatrix}0 & 1 & 0\\ & 0 & 1\\ h & & 0\end{pmatrix} = \begin{pmatrix}h & & \\ & h & \\ & & h\end{pmatrix}$. The second from the fact that if $h = N\gamma$ then $h^3 = \det h$ lies in $NE^\times$, as does $h^\ell$ and hence also $h$.

(3) $h = \alpha\begin{pmatrix}1 & & b\\ & a & \\ & & 1\end{pmatrix}$, $a \neq 1$, $b \neq 0$. Then $h = N\gamma$ if and only if $\alpha$ and $a$ lie in $NE^\times$. Indeed, since $\gamma$ belongs to $G_h(E)$ it takes the form $\gamma = \beta\begin{pmatrix}1 & & d\\ & c & \\ & & 1\end{pmatrix}$, and $N\gamma = N\beta\begin{pmatrix}1 & & \mathrm{tr}d\\ & Nc & \\ & & 1\end{pmatrix}$, so that $\alpha$ and $a$ lie in $NE^\times$.

(4) $h = \mathrm{diag}(\alpha,\beta,\gamma)$, $\beta \neq \alpha$. Then $h = N\gamma$ if and only if (a) $\beta$ lies in $NE^\times$, and (b) if $\ell \neq 2$ then $\alpha$ is in $NE^\times$. Indeed, if $h = N\gamma$ then

$\gamma = \begin{pmatrix} * & & * \\ & b & \\ * & & * \end{pmatrix}$ and $N\gamma = \begin{pmatrix} * & & * \\ & Nb & \\ * & & * \end{pmatrix}$ so that $\beta = Nb$. If $\ell \neq 2$ then

$\alpha^2 = \det\begin{pmatrix} \alpha & \\ & \alpha \end{pmatrix}$ is in $NE^\times$ as is $\alpha^\ell$, hence $\alpha$ lies in $NE^\times$. If $\ell = 2$

then $N\begin{pmatrix} & & 1 \\ & b & \\ \alpha & & \end{pmatrix} = \begin{pmatrix} \alpha & & \\ & \beta & \\ & & \alpha \end{pmatrix}$, hence no additional conditions are necessary.

(5) $h = \mathrm{diag}(a,b,c)$, $a,b,c$ distinct. Then $h = N\gamma$ if and only if $a$, $b$ and $c$ lie in $NE^\times$. This follows at once since such $\gamma$ must lie in $G_h(E)$ which is the diagonal subgroup.

(6) $h = \begin{pmatrix} \alpha & \\ & c \end{pmatrix}$ with a $GL_2(F)$-element $\alpha$ which lies in a quadratic torus $T(F)$ of $GL_2(F)$, and $c$ in $F^\times$. Then $h = N\gamma$ if and only if (a) $c$ is in $NE^\times$, and (b) $\alpha$ lies in $N_{T(E)/T(F)}T(E)$ if $T$ does not split over $E$, so that $T(E)$ is a field. Indeed, if $h = N\gamma$ then $\gamma = \begin{pmatrix} \beta & \\ & b \end{pmatrix}$ with $\beta$ in $T(E)$ and $b$ in $E^\times$. Hence the statement (a), and since $\alpha = N\beta$, we get also (b). If $T$ splits over $E$ and we denote the eigenvalues of $\alpha$ by $\alpha_1, \alpha_2 = \alpha_1^\sigma$, then $\begin{pmatrix} \alpha & \\ & c \end{pmatrix}$ is conjugate over $G(E)$ to $N\begin{pmatrix} \beta & \\ & b \end{pmatrix}$ with $\beta = \begin{pmatrix} \cdot & 1 \\ \alpha_1 & \end{pmatrix}$ and $b$ with $c = Nb$.

From local class field theory it follows that (b) can be replaced by the requirement that $\det \alpha$ lies in $NE^\times$, when $F$ is a local field.

(7) $h$ lies in a cubic torus $T$ of $G(F)$. Then $h = N\gamma$ if and only if $h$ lies in $N_{T(E)/T(F)}T(E)$ whenever $T$ does not split over $E$. Indeed if $h = N\gamma$ then $\gamma$ lies in the centralizer of $h$ in $G(E)$ which is $T(E)$, and the condition is obvious. If $T$ splits over $E$ we denote one of the eigenvalues of $h$ by $s$ and note that $h$ is conjugate over $G(E)$ to

$$N\begin{pmatrix} 0 & 1 & 0 \\ & 0 & 1 \\ s & & 0 \end{pmatrix} = \mathrm{diag}(s, s^\sigma, s^{\sigma^2}).$$

Using local class field theory we see that an equivalent condition is given by the requirement that det h lies in $NE^{\times}$, when $F$ is a local field.

1.1.3 Local v. global

We shall also need a description of the global map $N$ in terms of the local maps. If $F$ is a global field and $E$ is a cyclic extension of prime degree $\ell$ then we set $E_v = E \otimes_F F_v$ for each place $v$ of $F$. $G = G(E_v/F_v)$ acts on $E_v$. Either $E_v$ is a field and $G = G(E_v/F_v)$, or $E_v$ is isomorphic to a direct sum of $\ell$ copies of $F_v$ and $\sigma$ acts by mapping $(\gamma_1,\ldots,\gamma_\ell)$ to $(\gamma_2,\ldots,\gamma_\ell,\gamma_1)$. In the latter case $G(E_v)$ is isomorphic to a direct product of $\ell$ copies of $G(F_v)$, in which $G(F_v)$ is embedded diagonally. At a place which splits in $E$ each element is a norm since $N\gamma$ for $\gamma = (h,1,\ldots,1)$ is equal to the element $h = (h,h,\ldots,h)$ of $G(F_v)$. Thus we have:

LEMMA 1. *Suppose* $h$ *lies in* $G(F)$ *where* $F$ *is a global field. Then the equation* $h = N\gamma$ *has a solution* $\gamma$ *in* $G(E)$ *if and only if it has a solution in* $G(E_v)$ *for each place* $v$, *and if and only if it has a solution in* $G(E_v)$ *for all* $v$ *but one.*

*Proof.* It suffices to show that $h \doteq N\gamma$ admits a global solution if it has a local solution everywhere. This follows from the fact that $a$ in $F^{\times}$ lies in $NE^{\times}$ if and only if it lies in $NE_v^{\times}$ for all $v$, and from cases (1)-(5) above if the eigenvalues of $h$ lie in $F^{\times}$, from case (6) if $h$ is quadratic and from case (7) if $h$ is cubic. The last equivalence follows from the product formula for the global norm residue symbol.

1.1.4 Galois cohomology

Let $h$ be an element of $G(F)$, and denote by $G_h(F)$ the centralizer of $h$ in $G(F)$, namely the set of $g$ in $G(F)$ with $g^{-1}hg = h$. If $\gamma$ is an element in $G(E)$ then we can define the $\sigma$-centralizer $G_\gamma^\sigma(E)$ of $\gamma$ in $G(E)$ to be the set of $g$ in $G(E)$ with $g^{-\sigma}\gamma g = \gamma$. Since below we need to relate measure on $G_\gamma^\sigma(E)$ and on $G_h(F)$, $F$ local, for $h$ conjugate to $N\gamma$, we shall now recall some elementary facts of galois cohomology.

Suppose $\Gamma$ is a group which operates on a group $A$ as a group of automorphisms so that ${}^{st}a = {}^{s}({}^{t}a)$, where the image of $a$ under $s$ is denoted by ${}^{s}a$ ($s,t$ in $\Gamma$, $a$ in $A$). Denote by $Z^1(\Gamma,A)$ the set of functions (cocycles) $z$ from $\Gamma$ to $A$ such that $z(st) = {}^{s}z(t)z(s)$. Since $A$ acts on $Z^1(\Gamma,A)$ by $a: z \longrightarrow z_a$, where $z_a(s) = {}^{s}az(s)a^{-1}$, we have an equivalence relation $R$. The quotient of $Z^1(\Gamma,A)$ by $R$ is denoted by $H^1(\Gamma,A)$. The class of the cocycle which maps each $s$ to the identity of $A$ is denoted by $0$.

Let $X$ be a $\Gamma$-set on which $A$ acts as a transformations group, so that ${}^{s}(ax) = {}^{s}z\,{}^{s}x$ ($s$ in $\Gamma$, $a$ in $A$, $x$ in $X$; $ax$ denotes the image of $x$ by $A$). For any $z$ in $Z^1(\Gamma,A)$ we can get a new action of $\Gamma$ on $X$ by twisting with $z$. That is we put ${}_{s}x = z(s)^{-1}({}^{s}x)$ for any $s$ in $\Gamma$ and $x$ in $X$, and we have ${}_{st}x = {}_{x}({}_{t}x)$ since

$$z(st)^{-1}({}^{st}x) = (z(s)^{-1}\,{}^{s}z(t)^{-1})({}^{s}({}^{t}(x))$$

$$= z(s)^{-1}({}^{s}(z(t)^{-1}({}^{t}x))).$$

Thus $X$ is made into a new $\Gamma$-set $X_z$ (which has the same elements as $X$ but the action of $\Gamma$ is different). Note that if we put $_s a = z(s)^{-1}\, {}^s a z(s)$ then $_s(ax) = {_s a}\, {_s x}$ .

A particular case is obtained on replacing $X$ by $A$ and $A$ by the group $\operatorname{Aut} A$ of automorphisms of $A$ , in the above scheme. Here $A$ is a $\Gamma$-group, and $\Gamma$ operates an $\operatorname{Aut} A$ by $s: \alpha \longrightarrow {}^s\alpha$ , where ${}^s\alpha(a) = {}^s(\alpha({}^{s^{-1}}a))$ ($s$ in $\Gamma$, $\alpha$ in $\operatorname{Aut} A$ , $a$ in $A$). Clearly ${}^s(\alpha(a)) = {}^s\alpha({}^s a)$. As above, each $z'$ in $Z^1(\Gamma,\operatorname{Aut} A)$ defines a new $\Gamma$-group $A_{z'}$.

Let $z$ be in $Z^1(\Gamma,A/Z)$ where $A/Z$ denotes the quotient of $A$ by its centre $Z$ , and put $z' = \operatorname{ad} z$ , where $\operatorname{ad} z(s)$ is the automorphisms of $A$ which maps $a$ to $z(s)^{-1}az(s)$. We have $\operatorname{ad}\delta \cdot \operatorname{ad}\varepsilon = \operatorname{ad}\delta\varepsilon$ ($\delta,\varepsilon$ in $A$) and $\operatorname{ad}{}^s z(t) = {}^s(z'(t))$. Indeed,

$$\begin{aligned} {}^s(z'(t))(a) &= {}^s(\operatorname{ad} z(t))\,(a) = {}^s(\operatorname{ad} z(t)\,({}^{s^{-1}}a)) \\ &= {}^s(z(t)^{-1}\, {}^{s^{-1}}a\, z(t)) = {}^s z(t)^{-1}\, a\, {}^s z(t) \\ &= (\operatorname{ad}{}^s z(t))(a)\ . \end{aligned}$$

Hence

$$\begin{aligned} z'(st) &= \operatorname{ad} z(st) = \operatorname{ad}({}^s z(t) z(s)) = \operatorname{ad}({}^s z(t))\operatorname{ad} z(s) \\ &= {}^s z'(t) z'(s)\ , \end{aligned}$$

and we see that $z'$ gives an element of $H^1(\Gamma,\operatorname{Aut} A)$. We write $A_z$ for the $\Gamma$-group $A_{z'}$.

We shall be interested in the special case where $A$ is $G_h(E)$ for $h = N\gamma$ in $G(F)$, and where $\Gamma = G$ acts on $G_h(E)$ componentwise ($F$ local). Since $G$ is commutative we may write $a^\sigma$ for the image of $a$ (in $G_h(E)$) under $\sigma$ (in $G$). Since $G$ is cyclic each $\tau$ in $G$ can be put in the form $\tau = \sigma^r$ for some $r$. If $Z$ denotes the centre of $G_h(E)$ we put $z(\tau) = \gamma^{\sigma^{r-1}} \cdots \gamma^\sigma \gamma$ modulo $Z$, and as above $z' = \mathrm{ad}\, z$ defines an element of $H^1(G, \mathrm{Aut}\, G_h(E))$. The new $G$-group $A_z$ is denoted by $G_h^\sigma(E)$. We denote by $G_h^\sigma(F)$, or $A_z^G$, the group of $G$-invariant elements of $G_h^\sigma(E)$. It consists of all $g$ in $G_h(E)$ with $g = {}_\sigma g$, namely $g = \gamma^{-1} g^\sigma \gamma$, since

$$ {}_\sigma g = z'(\sigma)^{-1}({}^\sigma g) = z(\sigma)^{-1} g^\sigma z(\sigma) = \gamma^{-1} g^\sigma \gamma . $$

Hence we have $G_h^\sigma(F) = G_\gamma^\sigma(E)$; this fact will afford relating measures on $G_h(F)$ and on $G_\gamma^\sigma(E)$. Note that $G_h(F)$ is $A^G$, the set of $G$-invariant elements of $A$, consisting of the $g$ in $G_h(E)$ with $g = g^\sigma$.

Let $M$ be the algebra of $3 \times 3$ matrices and $M_h$ the centralizer of $h$ in $M$. The above cocycle $z$ defines a $G$-set $M_h^\sigma(E)$ whose multiplicative group is $G_h^\sigma(E)$. The $F$-algebras $A$ and $A'$ are said to be $E/F$-forms if they become isomorphic upon extension to $E$. The map $z' \longrightarrow A_{z'}$ is a bijection from $H^1(G, \mathrm{Aut}\, A)$ to set the set of isomorphism classes of $E/F$-forms of $A$; [14], X, §5, Prop. 8. We take $A = M_h$ and assume that $h = N\gamma$ is a scalar in $Z(F)$. Then $M_h = M$ and the automorphisms of $M$ are all of the form $\mathrm{ad}\, g$ with $g$ in $G/Z$. Hence $z \longrightarrow A_z = A_{z'}$ defines a bijection from $H^1(G, G/Z)$ to the set of isomorphism classes of $E/F$-forms of $A$, in this case. In particular $M_h^\sigma(F)$ is isomorphic to $M_h(F)$ if and only if $z'$ is equivalent to the trivial cocycle $0$. In this case $z'(\sigma) = (\text{ad } a)^{-\sigma}\text{ad } a$ for some $a$ in $G_h(E)$, hence $z(\sigma) = a^{-\sigma}a$ modulo $Z(E)$, and so $\gamma = a^{-\sigma}da$ for some scalar $d$ in $Z(E)$. Since all $E/F$-forms of $M_h(F)$ are isomorphic either to $M_h(F)$ or to a division algebra of dimension 9 over $F$ we deduce that if $h = N\gamma$ lies in $Z(F)$ and $\gamma$ is not $\sigma$-conjugate to a scalar then $M_h^\sigma(F)$ is a division algebra of dimension 9 over $F$ and $G_h^\sigma(F)$ is its multiplicative group.

As noted above it can be deduced that if $N\gamma$ and $N\delta$ are conjugate then $\gamma$ and $\delta$ are $\sigma$-conjugate. Indeed we may assume that $N\gamma = N\delta$ lies in $G(F)$. For $\tau = \sigma^r$ in $G$ we put

$$z(\tau) = \gamma^{-1}\gamma^{-\sigma}\cdots\gamma^{-\sigma^{r-1}}\delta^{\sigma^{r-1}}\cdots\delta^{\sigma}\delta .$$

Since $N\gamma = N\delta$, $z(1) = z(\sigma^{\ell})$ is equal to $1$ and $z(\tau)$ is well-defined. We have

$$z(\tau\sigma) = \gamma^{-1}z(\tau)^{\sigma}\gamma z(\sigma) = {}_{\sigma}z(\tau)z(\sigma) ,$$

so that $\tau \longrightarrow z(\tau)$ defines an element of $H^1(G,G_h^\sigma(E))$. Since $H^1(G,G_h^\sigma(E))$ is $\{0\}$ ([14], Chap. X, §1, Ex. 2), $z$ is trivial, that is, for some $a$ in $G_h^\sigma(E)$ we have $z(\sigma) = {}_{\sigma}a^{-1}a = \gamma^{-1}a^{\sigma}\gamma a$. But $z(\sigma) = \gamma^{-1}\delta$ and so $\delta = a^{-\sigma}\gamma a$, as required.

1.2.1 G(F)-families

Let $f$ be a compactly supported smooth (that is K-finite and also highly differentiable in the archimedean case) function on $G(F)$ ($F$ local). For each regular element $h$ in $G(F)$ the centralizer $G_h(F)$ of $h$ in $G(F)$ is a torus $T(F)$

in $G(F)$. Let $\omega_G$ and $\omega_T$ be invariant holomorphic forms of maximal degree on $G(F)$ and $T(F)$. They define measures on $G(F)$ and $T(F)$ whose quotient is denoted by $dg$. Put

$$F_0(h,f) = \int_{G_h(F)\backslash G(F)} f(g^{-1}hg)dg \quad (h \text{ regular in } G(F)).$$

Its dependence on $\omega_T$, $\omega_G$ will be indicated by writing $F_0(h,f;\omega_T,\omega_G)$.

A family $\{F_0(h);\ h \text{ regular in } G(F)\}$ of complex numbers will be called a $G(F)$-<u>family</u> if there exists a compactly supported smooth function $f$ on $G(F)$ such that $F_0(h) = F_0(h,f)$. We write $F_0(h;\omega_T,\omega_G)$ to specify the dependence of $F_0(h)$ on $\omega_T$ and $\omega_G$. Clearly a $G(F)$-family satisfies

(1) If $\omega_T' = a\omega_T$ and $\omega_G' = b\omega_G$ with $a,b$ in $F^\times$ then

$$F_0(h;\omega_T',\omega_G') = \left|\frac{b}{a}\right| F_0(h;\omega_T,\omega_G).$$

(2) If $T' = g^{-1}Tg$ and $h' = g^{-1}hg$ for $g$ in $G(F)$ and $\omega_{T'}$ is obtained from $\omega_T$ then

$$F_0(h';\omega_{T'},\omega_G) = F_0(h;\omega_T,\omega_G).$$

We can now record the classification of $G(F)$-families in the non-archimedean case, which was deduced from Howe's conjecture in Flath [4] and can be deduced also from Kottwitz [11]. For the proof of Proposition 5.5 in the event that for some place $v$ we have $F_v = \mathbb{R}$, $E_v = \mathbb{C}$, we are reduced to assuming that the following lemma is valid also in the archimedean case $(F = \mathbb{R}, E = \mathbb{C})$ although it successfully resisted all attempts to supply it with a proof.

LEMMA 2. *A set* $\{F_0(h);\ h$ *regular in* $G(F)\}$ *is a* $G(F)$-*family if and only if in addition to* (1) *and* (2) *above it satisfies the following*:

(3) *The restriction of* $h \longrightarrow F_0(h)$ *to each torus* $T(F)$ *is a smooth function on the set of regular elements in* $T(F)$ *and its support is relatively compact in* $T(F)$.

(4) *There exist functions* $F_1, F_2$ *smooth on* $F^\times \times F^\times$ *and* $F_3$ *smooth on* $F^\times$ *such that*: (i) *for each* $z$ *in* $F^\times$

$$F_0(h;\omega_T,\omega_G) = F_1(z,z)c_1(h;\omega_T,\omega_G) + F_2(z,z)c_2(h;\omega_T,\omega_G) + F_3(z)c_3(h;\omega_T,\omega_G)$$

*in a neighborhood of* $\mathrm{diag}(z,z,z)$ *in* $T(F)$, *for all* $T$ ; (ii) *for each* $x$ *and* $z$ *in* $F^\times$ *with* $x \neq z$

$$F_0(h;\omega_T,\omega_G) = F_1(x,z)c_1'(h;\omega_T,\omega_G) + F_2(x,z)c_2'(h;\omega_T,\omega_G)$$

*in a neighborhood of* $\mathrm{diag}(x,x,z)$ *in* $T(F)$ *for all split and quadratic* $T$ .

Here for each $T$ the functions $c_1, c_1', c_2, c_2', c_3$ are smooth functions on the regular subset of $T(F)$, independent of the set $\{F_0(h)\}$ and can be specified by the "necessary" direction of the Lemma. In particular

$$c_3(h;\omega_T,\omega_G) = \begin{cases} |T(F)\backslash G'(F)|\,, & \text{if } T \text{ is a cubic torus,} \\ 0\,, & \text{otherwise,} \end{cases}$$

where $G'(F)$ denotes the group of invertible elements of a division algebra of dimension 9 over $F$ , and the volume is taken with respect to $\omega_T$ and the measure $\omega_{G'}$ on the form $G'(F)$ of $G(F)$ which is obtained from $\omega_G$ in the usual way ([8],p. 475-6). Moreover $c_2$ and $c_2'$ vanish on split tori.

For a $G(F)$-family $F_0(h,f)$ we have

$$F_1(x,z) = \iiiint f\left(k^{-1}\begin{pmatrix} x & n_1 & n_2 \\ & x & n_3 \\ 0 & & z \end{pmatrix} k\right) dk\,dn_1\,dn_2\,dn_3 \,,$$

$$F_3(z) = f(\mathrm{diag}(z,z,z))$$

and

$$F_2(x,z) = \iiint f\left(k^{-1}\begin{pmatrix} x & 0 & n_2 \\ & x & n_3 \\ 0 & & z \end{pmatrix} k\right) dk\,dn_2\,dn_3 \quad (x,z \text{ in } F^\times).$$

1.2.2 Twisted $G(F)$-families

In addition to the orbital integrals $F_0(h,f)$ we consider, for any smooth compactly supported function $\phi$ on $G(E)$, the twisted orbital integrals

$$F_0(h,\phi) = \int_{G_\gamma^\sigma(E)\backslash G(E)} \phi(g^{-\sigma}\gamma g)\,dg \quad (\gamma \text{ in } G(E)).$$

Here $h$ denotes an element of $G(F)$ which is conjugate (in $G(E)$) to $N\gamma$. The notation on the left is justified since the integral on the right depends only on $h = N\gamma$ but not on $\gamma$ itself.

If $h$ lies in $G(F)$, then, as above, $G_\gamma^\sigma(E) = G_h^\sigma(F)$ is a form of $G_h(F)$, and measures on $G_h(F)$ can be transformed to $G_\gamma^\sigma(E)$ in the usual way

([8],pp. 475-6). We choose the corresponding measure on $G_\gamma^\sigma(E)$ . In particular, if $h$ is regular in $G(F)$ then $G_\gamma^\sigma(E) = G_h(F)$ is a torus $T(F)$ of $G(F)$ and we choose on it the measure defined by $\omega_T$. To specify the dependence on $\omega_G$ and $\omega_T$ from which $dg$ was obtained we write

$$F_0(h,\phi;\omega_T,\omega_G) \quad \text{for} \quad F_0(h,\phi) \quad (h = N\gamma \text{ regular in } G(F)).$$

A family $\{F_0(h);\ h \text{ regular in } G(F)\}$ of complex numbers is called a _twisted_ $G(F)$-_family_ if $F_0(h) = 0$ whenever there is no $\gamma$ in $G(E)$ with $h = N\gamma$ and if there exists $\phi$ as above such that for all $h$ with $h = N\gamma$ we have

$$F_0(h;\omega_T,\omega_G) = F_0(h,\phi;\omega_T,\omega_G) .$$

We are interested in the relations between the $G(F)$-families and the twisted $G(F)$-families. This is established in the following analogue of [12], Lemma 6.2.

1.2.3 _Matching orbital integrals_

LEMMA 3. _A twisted_ $G(F)$-_family is a_ $G(F)$-_family_. _A_ $G(F)$-_family_ $\{F_0(h)\}$ _is a twisted_ $G(F)$-_family if and only if_ $F_0(h) = 0$ _for all_ $h$ _not of the form_ $N\gamma$ _for any_ $\gamma$ _in_ $G(E)$. _Moreover, for a twisted_ $G(F)$-_family_ $\{F_0(h,\phi)\}$, _if_ $T$ _is cubic then_ $F_3(z)$ _is_ $0$ _unless there is_ $\gamma$ _in_ $G(E)$ _so that_ $z = N\gamma$ _when it is given by_

$$F_3(z) = \int_{G_\gamma^\sigma(E)\backslash G(E)} \phi(g^{-\sigma}\gamma g)dg.$$

<u>Proof.</u> A twisted $G(F)$-family clearly satisfies (1), (2). It suffices to verify (3) and (4) for $\phi$ which is supported on a small neighborhood of a given $\gamma$. If $N\gamma$ is regular (3) follows on considering the map $(t,g) \longrightarrow g^{-\sigma}tg$ ($t$ in the $\sigma$-regular set $T'(E)$ of the torus $T(E) = G^{\sigma}_{\gamma}(E)$, $g$ in $G^{\sigma}_{\gamma}(E)\backslash G(E)$) which realizes the set of $\sigma$-conjugates of $T'(E)$ in $G(E)$ as the quotient of $T'(E) \times T(F)\backslash G(E)$ by the Weyl group of $T$ in $G$ (under the action $w(t,g) = (w^{-\sigma}tw, w^{-1}g)$). We have to establish (4) for the semi-simple singular $N\gamma$. There are several cases to be dealt with.

Suppose $\gamma$ is $\sigma$-conjugate to a central element. Conjugating we may assume that $\gamma$ is central and translating we may suppose that $\gamma$ is 1. We choose an analytic section $s$ of $G'(F) \longrightarrow G(F)\backslash G'(F)$, where $G' = \mathrm{Res}_{E/F}G$ is the group obtained from $G$ by restriction of scalars from $E$ to $F$ ; note that $G'(F) = G(E)$. The map

$$G(F) \times G(F)\backslash G'(F) \longrightarrow G'(F) \quad \text{defined by } (g,w) \longrightarrow s(w)^{-\sigma}gs(w)$$

gives an analytic isomorphism in a neighborhood of $1$ in $G'(F)$.

If $\phi$ is supported on such neighborhood and $\varepsilon$ lies in its intersection with $G(F)$ we put

$$f(\varepsilon) = \int_{G(F)\backslash G'(F)} \phi(s(w)^{-\sigma}\varepsilon s(w)).$$

Since

$$\int_{G^{\sigma}_{\varepsilon}(E)\backslash G(E)} \phi(g^{-\sigma}\varepsilon g) = \int_{G_{\varepsilon}(F)\backslash G(F)}\int_{G(F)\backslash G'(F)} \phi(s(w)^{-\sigma}g^{-1}\varepsilon g s(w))$$
$$= \int_{G_{\varepsilon}(F)\backslash G(F)} f(g^{-1}\varepsilon g)$$

we have

$$F_0(\varepsilon^\ell,\phi) = F_0(\varepsilon,f) .$$

Since extraction of the $\ell$-th root of unity is a (single-valued) function in a sufficiently small neighborhood of 1 (4) follows in this case, together with the expression for $F_3(z)$ in the lemma.

Suppose that $\gamma$ is $\sigma$-conjugate to a scalar multiple of some $\mathrm{diag}(1,1,\alpha)$ with $N\alpha \neq 1$. Conjugating and translating we assume that $\gamma = \mathrm{diag}(1,1,\alpha)$. We choose an analytic section $s$ of $G'(F) \longrightarrow M_1(F)\backslash G'(F)$, and note that the map

$$M_1(F) \times M_1(F)\backslash G'(F) \longrightarrow G'(F) \quad \text{defined by} \quad (g,w) \longmapsto s(w)^{-\sigma} g s(w)$$

gives an analytic isomorphism in a neighborhood of $\gamma$. Note that if an element of $G(F)$ is close to $N\gamma$ it must lie in a quadratic or split torus since $N\alpha \neq 1$, and conjugating we may assume that it lies in a torus of $M_1(F)$.

Let $\phi$ be a smooth function supported on a small neighborhood of $\gamma$ as above and suppose $\varepsilon$ in $M_1(F)$ is so close to 1 that $\phi(\varepsilon\gamma) \neq 0$. Put

$$f_\gamma(\varepsilon) = \int_{M_1(F)\backslash G'(F)} \phi(s(w)^{-\sigma}\gamma\varepsilon s(w)).$$

Then

$$F_0(\varepsilon^{\ell}N\gamma,\phi) = \int_{G^{\sigma}_{\varepsilon\gamma}(E)\backslash G(E)} \phi(g^{-\sigma}\varepsilon\gamma g)$$

$$= \int_{M_{1\varepsilon}(F)\backslash M_1(F)} \int_{M_1(F)\backslash G'(F)} \phi(s(w)^{-\sigma}g^{-1}\varepsilon\gamma g s(w))$$

$$= \int_{M_{1\varepsilon}(F)\backslash M_1(F)} f_{\gamma}(g^{-1}\varepsilon g)dg = F_0(\varepsilon,f_{\gamma}).$$

The family $\{F_0(\varepsilon,f_{\gamma})\}$ is an $M_1(F)$-family, which is the same as GL(2,F)-family. Its asymptotic behaviour is given in [12], Lemma 6.1, and (up to a multiple depending on the eigenvalues of $\varepsilon$ and $\gamma$, which can be introduced in the definition of $f_{\gamma}$) it is of the type described in (4) (ii), since any element in $M_1(F)$ sufficiently close to $N\gamma$ in $M_1(F)$ can be expressed in the form $\varepsilon^{\ell}N\gamma$ for a unique $\varepsilon$ in a sufficiently small neighborhood of 1 in $M_1(F)$.

The next case to be considered is when $h = N\gamma$ is central in $G(F)$ but $\gamma$ is not $\sigma$-conjugate to a central element. Then $\ell = 3$ and, as we saw above, we obtain a twisted form $G^{\sigma}_h(E)$ of $G_h(E) = G(E)$, with $G^{\sigma}_h(F) = G^{\sigma}_{\gamma}(E)$. Moreover since $\gamma$ is not $\sigma$-conjugate to a scalar we have that $G^{\sigma}_{\gamma}(E)$ is the group of invertible elements in a division algebra of dimension 9 over $F$. For $\varepsilon$ in $G^{\sigma}_{\gamma}(E)$ sufficiently close to 1 we have $G^{\sigma}_{\gamma\varepsilon}(E) \subset G^{\sigma}_{\gamma}(E)$ since $G^{\sigma}_{\gamma\varepsilon}(E) \subset G_{\varepsilon}(E)$. We take a section $s$ of $G'(F) \longrightarrow G^{\sigma}_{\gamma}(E)\backslash G'(F)$. Suppose $\phi$ is supported on a small neighborhood of $\gamma$ in $G'(F) = G(E)$ and $\varepsilon$ in $G^{\sigma}_{\gamma}(E)$ is such that $\gamma\varepsilon$ lies there. We put

$$f(\varepsilon) = \int_{G_\gamma^\sigma(E)\backslash G'(F)} \phi(s(w)^{-\sigma}\gamma\varepsilon s(w)) ,$$

and then

$$f_0(z\varepsilon^3,\phi) = \int_{G_{\gamma\varepsilon}^\sigma(E)\backslash G'(F)} \phi(g^{-\sigma}\gamma\varepsilon g)$$

$$= \int_{G_{\gamma\varepsilon}^\sigma(E)\backslash G_\gamma^\sigma(E)} \int_{G_\gamma^\sigma(E)\backslash G'(F)} \phi(s(w)^{-\sigma}\gamma g^{-1}\varepsilon g s(w))$$

$$= \int_{G_\varepsilon(E)\cap G_\gamma^\sigma(E)\backslash G_\gamma^\sigma(E)} f(g^{-1}\varepsilon g) = F_0(\varepsilon,f).$$

Here $\{F_0(\varepsilon,f)\}$ is a $G_\gamma^\sigma(E) = G_h^\sigma(F)$-family. Since this is the group of invertible elements in a division algebra the statement (4)(i) is easy to establish, in fact with $c_1 = c_2 = 0$ and the correct value of $c_3$ on cubic tori. Note that $G_\gamma^\sigma(E)$ does not have split or quadratic tori. The required expression for $F_3(z)$ follows from the expression for $f$ given above.

The remaining case is when $N\gamma = h = \mathrm{diag}(x,x,N\alpha)$, where $\alpha$ in $E^\times$ is with $N\alpha \neq x$, and $\gamma$ is not $\sigma$-conjugate to any diagonal element. Then $\ell = 2$. The map $z' = \mathrm{ad}\, z$, with $z(\sigma) = \gamma$, defines an element of $H^1(G,\mathrm{Aut}\, G_h(E))$, and hence a twisted form $G_h^\sigma(E)$ of $G_h(E) = M_1(E)$ with $G_h^\sigma(F) = G_\gamma^\sigma(E)$. The form $G_h^\sigma(F)$ is equal to $M_1(F)$ if and only if this cocycle is cohomologous to the trivial cocycle, that is if $a^{-\sigma}\gamma a$ lies in $A_1(E)$ for some $a$ in $G(E)$. But $\gamma$ is not $\sigma$-conjugate to any element of $A_1(E)$ and so $G_\gamma^\sigma(E)$ is the direct product of the group of invertible elements in a division algebra of dimension 4 over $F$, and of $GL(1,F) = F^{\times}$. As before we note that for $\varepsilon$ in $G_{\gamma}^{\sigma}(E)$ near 1 we have $G_{\gamma\varepsilon}^{\sigma}(E) \subset G_{\gamma}^{\sigma}(E)$ since $G_{\gamma\varepsilon}^{\sigma}(E) \subset G_{\varepsilon}(E)$. So we choose a section $s$ of $G'(F) \longrightarrow G_{\gamma}^{\sigma}(E)\backslash G'(F)$ and a smooth $\phi$ with support in a small neighborhood of $\gamma$ in $G'(F)$. For $\varepsilon$ in $G_{\gamma}^{\sigma}(E)$ with $\phi(\gamma\varepsilon) \neq 0$ we put

$$f_{\gamma}(\varepsilon) = \int_{G_{\gamma}^{\sigma}(E)\backslash G'(F)} \phi(s(w)^{-\sigma}\gamma\varepsilon s(w))$$

so that

$$F_0(z\varepsilon^2,\phi) = F_0(\varepsilon,f_{\gamma}) .$$

Here $\{F_0(\varepsilon,f_{\gamma})\}$ is a $G_{\gamma}^{\sigma}(E)$-family and (4) (ii) is easy to establish since $G_{\gamma}^{\sigma}(E)$ is the direct product of the group of invertible elements in a division algebra and of $GL(1,F)$. In fact we have $c_1' = 0$, $G_{\gamma}^{\sigma}(E)$ has only quadratic tori for which $c_2'$ is non-zero, and up to a multiple (which depends only on the eigenvalues of $\varepsilon$ and $\gamma$ and so it can be incorporated in the definition of $f$), it is the requested function.

### 1.2.4 End of Proof

We have proved the first and the last statements of Lemma 3. It remains to show that a $G(F)$-family $\{F_0(h)\}$ which vanishes at any $h$ not of the form $N\gamma$ is a twisted $G(F)$-family. As usual this claim is easy for $h$ on the regular set so that we can restrict our attention to a family supported on a small neighborhood of a singular element $z$. Suppose $z$ lies in $Z(F)$. If it lies in $NZ(E)$ we may assume that it is 1; otherwise $\ell = 3$. In both cases there is $h$ in $G(F)$ with $Nh = h^{\ell} = z$ , and every element in a small neighborhood of $z$ can be expressed in the form $\gamma^{\ell}$ with $\gamma$ in $G(F)$ sufficiently close to $h$ , for a unique $\gamma$. Hence it suffices to find $\phi$ such that $F_0(\gamma^{\ell},\phi) = F_0(\gamma^{\ell})$ , for $\gamma$ near $h$ . Consider $\{F_0'(\gamma) = F_0(\gamma^{\ell})\}$; it is a $G(F)$-family since it satisfies (1)-(4). Hence for some $f$ we have $F_0'(\gamma) = F_0'(\gamma,f)$. We may assume that the support of $f$ consists only of elements whose conjugates are near $h$ , and moreover, applying a partition of unity and conjugating we may assume that $f$ is supported on a small neighborhood of $h$ . If $\alpha$ is a function on $G(F)\backslash G'(F)$ whose integral on this space is 1 we define

$$\phi(s(w)^{-\sigma}gs(w)) = \alpha(w)f(g) \quad (g \text{ in } G(F),\ w \text{ in } G(F)\backslash G'(F)).$$

This function satisfies $F_0(\gamma^{\ell},\phi) - F_0(\gamma^{\ell})$ for $\gamma$ near $h$ , as required.

Finally, we consider a family $\{F_0(h)\}$ which is supported on a small neighborhood of an element $N\gamma = \mathrm{diag}(x,x,N\alpha)$ with $x$ in $F^{\times}$ , $\alpha$ in $E^{\times}$, $N\alpha \neq x$. If $\delta$ is in $G(E)$ and $N\delta$ is near $N\gamma$ then $N\delta$ must lie in a quadratic or a split torus. Upon conjugating we may assume that $\delta$ lies in $M_1(E)$ . In particular $\gamma$ lies in $M_1(E)$ . Every element in $M_1(F)$ sufficiently close to $N\gamma$ can be written as $\delta^{\ell}N\varepsilon$ ($\varepsilon = \mathrm{diag}(1,1,\alpha)$) for a unique $\delta$ in $M_1(F)$ such that $\delta\varepsilon$ is in a sufficiently small neighborhood of $\gamma$ . In particular we may assume that $\gamma\varepsilon^{-1}$ lies in $M_1(F)$ . So it suffices to find $\phi$ such that $F_0(\delta^{\ell}N\varepsilon,\phi)$ is equal to $F_0(\delta^{\ell}N\varepsilon)$ for $\delta$ in such a small neighborhood of $\gamma\varepsilon^{-1}$ in $M_1(F)$ .

The family $\{F_0(\delta) = F_0(\delta^\ell N\varepsilon)\}$ is an $M_1(F)$-family since it satisfies (1)-(4). Hence for some $f$ on $M_1(F)$ we have $F_0(\delta,f) = F_0(\delta)$. As usual we may assume that $f$ is supported only on a small neighborhood of $\gamma\varepsilon^{-1}$ in $M_1(F)$. Let $s$ be a section of $G'(F) \longrightarrow M_1(F)\backslash G'(F)$, and let $\alpha$ be a function on $M_1(F)\backslash G'(F)$ whose integral over the space is $1$. Put

$$\phi(s(w)^{-\sigma}\varepsilon g s(w)) = \alpha(w)f(g) \quad (g \in M_1(F),\ w \in M_1(F)\backslash G'(F))$$

and note that

$$F_0(\delta^\ell N\varepsilon,\phi) = F_0(\delta,f) = F_0(\delta) = F_0(\delta^\ell N\varepsilon)$$

for $\delta$ in $M_1(F)$ near $\gamma\varepsilon^{-1}$. It follows that $F_0(\delta,\phi) = F_0(\delta)$ for all $\delta$ in $G(F)$, as required.

1.2.5 Reformulation

Let $\omega$ be a quasi-character of $NZ(E)$, and let $\omega_E$ be the quasi-character of $Z(E)$ with $\omega_E(x) = \omega(Nx)$ ($x$ in $E^\times$). The results of Lemma 3 remain valid if instead of the compactly supported functions $f$ and $\phi$ we consider a smooth $f$ which is compactly supported modulo $NZ(E)$ and transforms by

$$f(zg) = \omega(z)^{-1}f(g) \quad (z \text{ in } NZ(E))$$

on $NZ(E)$, and a smooth $\phi$, compactly supported modulo $Z(E)$ and transforming under $Z(E)$ by

$$\phi(zg) = \omega_E(z)^{-1}\phi(g) \quad (z \text{ in } Z(E)).$$

Note that in the sequel it will be more useful to consider not the family $\{F_0(h)\}$ but the normalized orbital integrals $\{F(h)\}$ which are defined (for any regular $h$ in $G(F)$) by

$$F(h,f) = \Delta(h)\int_{G_h(F)\backslash G(F)} f(g^{-1}hg) = \Delta(h)F_0(h,f)$$

and (for $\gamma$ in $G(E)$ so that $h = N\gamma$ is regular in $G(F)$) by

$$F(h,\phi) = \Delta(h)\int_{G_\gamma^\sigma(E)\backslash G(E)} \phi(g^{-\sigma}\gamma g) = \Delta(h)F_0(h,\phi).$$

Here for every regular element $h$ in $G(F)$ with eigenvalues $h_1$, $h_2$, $h_3$ we put

$$\Delta(h) = \left| \frac{h_1 - h_2}{h_3} \; \frac{h_1 - h_3}{h_1} \; \frac{h_2 - h_3}{h_2} \right| .$$

Lemma 3 implies that for every smooth $f$ which has the property that $F(h,f) = 0$ when $h$ is not a norm and which is compactly supported on $G(F)$ mod $NZ(E)$ and transforms under $NZ(E)$ by $\omega^{-1}$, there exists a smooth $\phi$ which is compactly supported on $G(E)$ mod $Z(E)$ and transforms under $Z(E)$ by $\omega_E^{-1}$, such that

$$F(h,\phi) = F(h,f)$$

for all regular $h$ in $G(F)$. Lemma 3 also implies that for all $\phi$ there exists $f$ for which the above identity is valid. We denote this correspondence by $\phi \longrightarrow f$.

1.2.6 Spherical functions

Let $H$ be the convolution algebra of complex valued spherical (left and right $K(F)$-invariant) functions on $G(F)$ which are compactly supported modulo $NZ(E)$ and transform under $NZ(E)$ by $\omega^{-1}$. Here we assume that $F$ is a non-archimedean local field, and that $\omega$ is unramified. For any $z = (s,t,r)$ in $\mathbb{C}^{\times 3}$ let $\eta_z$ be the unramified character of $P_0(F)$ whose value at a matrix in $P_0(F)$ with diagonal $(\tilde{\omega}^{m_1}, \tilde{\omega}^{m_2}, \tilde{\omega}^{m_3})$ is $s^{m_1} t^{m_2} r^{m_3}$. Here $\tilde{\omega}$ denotes the local uniformizing parameter of $F$.

The modular function $\delta_0$ of $P_0$ is defined by $d(p'p) = \delta_0(p')dp$ where $dp$ is the right Haar measure on $P_0(p,p'$ in $P_0(F))$. We put

$$\eta_z'(g) = \eta_z(p)\delta_0^{1/2}(p) \quad (g = pk,\ p \text{ in } P_0(F),\ k \text{ in } K(F))$$

where $g$ in $G(F)$ is expressed in the form $pk$ according to the Iwasawa decomposition $G = P_0K$.

The Satake transform $f^\wedge$ of $f$ in $H$ is defined by

$$f^\wedge(z) = \int_{NZ(E)\backslash G(F)} f(g)\eta_z'(g)dg$$

on the set of $z$ with $(str)^\ell = \omega(\tilde{\omega}^\ell)$ if $E$ is an unramified cyclic extension of degree $\ell$ over $F$. It gives an isomorphism of $H$ with the algebra of finite Laurent series in $s,t$ and $r$ which are invariant under the group of permutations of $s,t$ and $r$, and where $(str)^{\ell} = \omega(\tilde{\omega}^{\ell})$ if $E/F$ is unramified.

We write $H(F)$ for $H$ and for our unramified extension $E$ of $F$ we write $H(E)$ for the convolution algebra of complex valued spherical (with respect to $K(E)$) functions on $G(E)$ which are compactly supported modulo $Z(E)$ and transform under $Z(E)$ by $\omega_E^{-1}$. The Satake transform $\phi^\wedge$ of $\phi$ in $H(E)$ is defined as above on the set of $z = (s,t,r)$ in $\mathbb{C}^{\times 3}$ with $str = \omega_E(\tilde{\omega})$. Let $\phi$ be in $H(E)$, and define $f = b(\phi)$ to be the unique element of $H(F)$ with

$$b(\phi)^\wedge(z) = \phi^\wedge(s^{\ell},t^{\ell},r^{\ell}) \quad (z = (s,t,r)).$$

This is a $\mathbb{C}$-algebras homomorphism from $H(E)$ to $H(F)$. It satisfies:

LEMMA 4. <u>For every</u> $\phi$ <u>in</u> $H(E)$ <u>we have</u> $\phi \longrightarrow b(\phi)$.

This is Theorem 8.9 of Kottwitz [11].

In other words, for every regular $h$ in $G(F)$ we have $F(h,\phi) = F(h,f)$ where $f = b(\phi)$. This is the sense which will be attached to the map $\phi \longrightarrow f$ in the sequel for spherical $\phi$, namely if $\phi$ is spherical we shall write $\phi \longrightarrow f$ if and only if $f = b(\phi)$. In particular we have $F(h,\phi^0) = F(h,f^0)$, where $f^0$ (resp. $\phi^0$) is the function whose value at $g$ is $0$ unless $g = zk$ ($z$ in $NZ(E)$, $k$ in $K(F)$ (resp. $z$ in $Z(E)$, $k$ in $K(E)$)) where its value is the quotient of $\omega^{-1}(z)$ (resp. $\omega_E^{-1}(z)$) by the volume of $K(F)$ (resp. $K(E)$). Lemma 3 implies that for a scalar $z$ in $Z(F)$ we have

$$(b(\phi))(z) = \int_{G_\gamma^\sigma(E)\backslash G(E)} \phi(g^{-\sigma}\gamma g) \quad (z = N\gamma) .$$

From chapter 3 onwards we shall consider not only $\phi$ but also the function $\phi'$ on the semi-direct product $G \times G(E)$ whose value at $(\tau,g)$ is 0 unless $\tau = \sigma$ when it is $\phi(g)$. If $\phi \longrightarrow f$ then we shall also write $\phi' \longrightarrow f$ .

1.3.1 Classification

Let $F$ be a non-archimedean local field. We recall some well-known facts about the classification of admissible irreducible representations of $G(F)$. Any such representation is either supercuspidal or an irreducible constituent in the decomposition series of some induced representation $I_{P_1}$ (or $I_{P_2}$) from $P_1(F)$ (or $P_2(F)$). To describe the latter representations consider first the induced representations from $P_0$ to $G$ of the form $I_{P_0}(\eta)$ with $\eta = (\eta_1,\eta_2,\eta_3)$ and $\eta_i = \mu_i\alpha^{s_i}$, where $\mu_i$ is a unitary character and $s_i$ is in $\mathbb{R}$ $(1 \leq i \leq 3)$. We may assume that $s_1 \geq s_2 \geq s_3$ . Then $I_{P_0}(\eta)$ is irreducible unless some of $\eta_1/\eta_2$, $\eta_2/\eta_3$ or $\eta_1/\eta_3$ are equal to $\alpha$ .

Suppose $\eta = (\mu\alpha,\mu,\mu\alpha^{-1})$. Then $I_{P_0}(\eta)$ has a quotient which is the one-dimensional representation $\pi_{P_0}(\eta) = \mu\cdot\det$ . The kernel of the natural map $I_{P_0}(\eta) \longrightarrow \pi_{P_0}(\eta)$ has a composition series consisting of the reducible induced representation $I_{P_2}(\mu\alpha,\sigma(\mu,\mu\alpha^{-1}))$ and the irreducible $\pi_{P_1}(\sigma(\mu\alpha,\mu),\mu\alpha^{-1})$ (see below). Here $\sigma(\eta_i,\eta_j)$ denotes the special representation of $GL(2,F)$. If we have $\eta$ with $\eta_i/\eta_j = \alpha$ for a single pair $i > j$ $(i,j = 1,2,3)$ then $I_{P_0}(\eta)$ has an irreducible quotient $\pi_{P_0}(\eta) = I_{P_2}(\eta_k,\eta_j\alpha^{1/2}\circ\det)$ with $k \neq i,j$ $(k = 1,2,3)$. The kernel of the natural map $I_{P_0}(\eta) \longrightarrow \pi_{P_0}(\eta)$ is the irreducible representation $I_{P_2}(\eta_k,\sigma(\eta_j\alpha,\eta_j))$.

Since $I_{P_1}(\tau,\eta) = I_{P_2}(\eta,\tau)$ for representations $\eta$ and $\tau$ of $GL(1)$ and $GL(2)$ it suffices to consider $I_{P_1}(\tau \otimes \alpha^{s_1},\mu\alpha^{s_2})$ and $I_{P_2}(\mu\alpha^{s_1}, \tau \otimes \alpha^{s_2})$ where $\mu$ is unitary and $s_1 \geq s_2$. If $\tau$ is supercuspidal then both are irreducible. If $\tau$ is $\pi(\eta_1,\eta_2)$ the theorem of induction in stages reduces the case to the above description of $I_{P_0}(\eta)$ if $\eta_1/\eta_2 \neq \alpha$. Otherwise we have to note that $I_{P_1}(\mu\alpha^s\circ\det,\mu'\alpha^{s'})$ is irreducible unless $\mu = \mu'$ and $s = s' + \frac{3}{2}$. In this case the decomposition series consists of the one-dimensional irreducible quotient $\pi_{P_0}(\eta)$ and an irreducible subrepresentation $\pi_{P_1}(\sigma(\mu\alpha,\mu),\mu\alpha^{-1})$ (see below). It remains to consider the case where $\tau$ is $\sigma(\mu\alpha^{1/2},\mu\alpha^{-1/2})$; we may assume that $\mu$, and hence $\tau$, is unitary. Then $I_{P_2}(\mu'\alpha^{s'},\tau \otimes \alpha^s)$ is irreducible unless $\mu = \mu'$ and $s' = s + 3/2$. In this case it has an irreducible quotient $\pi_{P_2}(\mu'\alpha^{s'},\tau\otimes\alpha^s)$. The kernel of the natural map is $\sigma \otimes \mu\alpha^{s+1/2}$ where $\sigma = \sigma(\alpha,1,\alpha^{-1})$ is the (irreducible) Steinberg representation. Similarly $I_{P_1}(\tau\otimes\alpha^s,\mu'\alpha^{s'})$ is irreducible unless $\mu' = \mu$ and $s = s' + 3/2$ where the decomposition series consists of a quotient $\pi_{P_1}(\tau\otimes\alpha^s,\mu'\alpha^{s'})$ and a subrepresentation $\sigma \otimes \mu\alpha^{s'+1/2}$.

In addition to the above notations following the usual conventions for GL(2) we shall denote $I_P$ by $\pi_P$ whenever $I_P$ is irreducible. Note that the Steinberg representation $\sigma \otimes \eta$ is the unique (irreducible) square-integrable constituent in the decomposition series of $I_{P_0}(\alpha,1,\alpha^{-1}) \otimes \eta$ (for any quasi-character $\eta$).

1.3.2 Twisted characters

Let E be a cyclic extension of degree $\ell$ of a local field F , and let $\sigma$ be a fixed generator of the galois group $G = G(E/F)$. If $\pi^E$ is an admissible irreducible representation of G(E) such that $\pi^E$ is equivalent to ${}^\sigma\pi^E$ (where ${}^\sigma\pi^E(g) = \pi^E(g^\sigma)$) then ${}^\sigma\pi^E = A\pi^E A^{-1}$ for some operator A of order $\ell$ , and $\pi^E$ extends to a representation of the semi-direct product $G \times G(E)$ by $\pi^E(\sigma)v = Av$ for v in the space of $\pi^E$ . Any other extension is of the form $\zeta \otimes \pi^E$ where $\zeta$ is the character of $G$.

Suppose ${}^\sigma\pi^E \approx \pi^E$, and that $\pi^E$ transforms under Z(E) by a character $\omega_E$. Suppose $\phi'$ is a smooth function on $G \times G(E)$ which transforms under Z(E) by $\omega_E^{-1}$ and its support is compact modulo Z(E) . The operator

$$\pi^E(\phi') = \int_{Z(E)\backslash G\times G(E)} \phi'(g)\pi^E(g)dg$$

has finite rank and its "twisted" character $\operatorname{tr} \pi^E(\phi')$ is finite.

The character is said to be a function if there exists a function $\chi_{\pi^E}$ on $G \times G(E)$ which transforms on Z(E) by $\omega_E$ such that for all $\phi'$ we have

$$\operatorname{tr} \pi^E(\phi') = \int_{Z(E)\backslash G\times G(E)} \phi'(g)\chi_{\pi^E}(g)dg \ .$$

We shall need the result that the character exists as a locally integrable function on $G \times G(E)$ which is smooth on the $(\sigma',\gamma)$ such that $N\gamma = \gamma^{\sigma'^{\ell-1}} \ldots \gamma^{\sigma'}\gamma$ is regular. Since these properties are known on the subgroup $G(E)$ of $G \times G(E)$ and $\sigma$ is an arbitrary generator of $G$, it suffices to establish them on the subset $\sigma \times G(E)$. Special cases can be handled now but the proof of this will be complete only with Lemma 6.2 as a result of the global theory.

1.3.3 Induced representations

Let $(\tau,\eta)$ be an admissible representation of $M_1(E)$ with ${}^\sigma\eta = \eta$, ${}^\sigma\tau \simeq \tau$.* The representation of $G \times G(E)$ induced from $(\tau,\eta)$ on $G \times M_1(E)$ is the same as the representation of $G \times G(E)$ extended from the representation $I_{P_1}(\tau,\eta)$ of $G(E)$ which is in turn induced from the representation $(\tau,\eta)$ of $M_1(E)$. The representation of $G \times G(E)$ is denoted again by $I_{P_1}(\tau,\eta)$. We write $K_1$ for $K \cap M_1\backslash K$ and choose a Haar measure on $K_1$ so that

$$\int_{Z(E)\backslash G(E)} h(g)dg = \int_{K_1(E)}\int_{N_1(E)}\int_{Z(E)\backslash M_1(E)} h(mnk)dmdndk \ .$$

For any $(\varepsilon,k)$ in $G\times G(E)$ and $\psi$ in the space of $I_{P_1}(\tau,\eta)$ we have

---

*For any representation $\pi$ of a group $H$ defined over $E$, ${}^\sigma\pi$ is defined by ${}^\sigma\pi(h) = \pi(h^\sigma)$ ($h$ in $H(E)$).

$$(I_{P_1}(\tau,\eta;\phi')\psi)((\varepsilon,k)) = \int_{Z(E)\backslash G\times G(E)} \phi'((\sigma',g))\psi((\varepsilon,k)(\sigma',g))$$

$$= \int \phi'((\varepsilon,k)^{-1}(\sigma',g))\psi((\sigma',g)) = \int_{G\times K_1(E)} K((\varepsilon,k),(\sigma',k_1))\psi((\sigma',k_1))dk_1$$

where $K((\varepsilon,k),(\sigma',k_1))$ is defined by

$$\iint \phi'((\varepsilon,k)^{-1}(\varepsilon\sigma,1)\,(\sigma',mnk_1))\delta^E_{M_1}(m)^{1/2}(\tau,\eta)((\varepsilon\sigma,1)(1,m^{\sigma'^{-1}}))dmdn$$

and $m \in Z(E)\backslash M_1(E)$, $n \in N_1(E)$.

Suppose that $\phi'$ is supported on $\sigma \times G(E)$ and put $\phi(g) = \phi'((\sigma,g))$ ($g$ in $G(E)$). Hence the kernel is non-zero only when $\sigma' = 1$, and since on the diagonal $\varepsilon$ is equal to $\sigma'$, we have

$$tr\{I_{P_1}(\tau,\eta;\phi')\} = \int \delta^E_{M_1}(m)^{1/2}\chi_{(\tau,\eta)}((\sigma,m))\,\phi(k^{-\sigma}mnk)dndkdm \qquad (*)$$

($m$ in $Z(E)\backslash M_1(E)$, $k$ in $K_1(E)$, $n$ in $N_1(E)$). Here $\chi_{(\tau,\eta)}$ denotes the character of $(\tau,\eta)$ and $\delta^E_{M_1}(\gamma) = \delta_{M_1}(N\gamma)$ is the modular function of $M_1(E)$.

For any element $h$ in $GL(2,F)$ with eigenvalues $h_1, h_2$ we write $\Delta_1(h)$ for $|(h_1 - h_2)^2/h_1h_2|^{1/2}$. If $h$ lies in $M_1(F)$ we denote by $h'$ its projection on $GL(2,F)$ and put $\Delta_1(h) = \Delta_1(h')$. The twisted analogue of the Weyl integration formula for $\sigma \times GL(2,E)$ (described before Lemma 7.2 in [12]) implies that the right side of (*) is

$$\frac{1}{2}\sum_{T_1}\int \Delta_1(N\gamma)^2\{\int \delta^E_{M_1}(\gamma)^{1/2}\chi_{(\tau,\eta)}((\sigma,m^{-\sigma}\gamma m))\iint \phi(k^{-\sigma}m^{-\sigma}\gamma mnk)dndkdm\}d\gamma\ .$$

The first integral is taken over $\gamma$ in $Z(E)T_1(E)^{1-\sigma}\backslash T_1(E)$, and $n \in N_1(E)$, $k \in K_1(E)$. The second integral is taken over $m$ in $Z(E)T_1(F)\backslash M_1(E)$. The sum is taken over a set of representatives for the conjugacy classes of Cartan subgroups $T_1$ of $M_1$ over $F$.

If $h$ in $M_1(F)$ has the eigenvalues $h_1, h_2$ and $e$ (the first two are the eigenvalues of the projection $h'$ of $h$ to $GL(2,F)$) then $\delta_{M_1}(h)$ is equal to $|h_1h_2/e^2|$, and as usual we put

$$\Delta(h) = \left|\frac{h_1 - h_2}{e}\,\frac{h_1 - e}{h_1}\,\frac{h_2 - e}{h_2}\right| .$$

We have

$$\Delta_1(N\gamma)\delta^E_{M_1}(\gamma)^{1/2}\iiint\phi(k^{-\sigma}m^{-\sigma}\gamma mnk) = \Delta(N\gamma)\iiint\phi(k^{-\sigma}n^{-\sigma}m^{-\sigma}\gamma mnk) \qquad (**)$$

and

$$\chi_{(\tau,\eta)}((\sigma,m^{-\sigma}\gamma m)) = \chi_{(\tau,\eta)}((\sigma,\gamma)) \quad (m \text{ in } M_1(E)),$$

since the character of $(\tau,\eta)$ is a class function. It follows that $(*)$ is equal to

$$\frac{1}{2}\sum_{T_1}\int_{Z(E)T_1(E)^{1-\sigma}\backslash T_1(E)} \Delta(N\gamma)F(N\gamma,\phi)(\Delta_1(N\gamma)\ \chi_{(\sigma,\eta)}((\sigma,\gamma))/\Delta(N\gamma))d\gamma ,$$

where $F(N\gamma,\sigma)$ is defined by the right side of $(**)$. (The notation is justified since $(**)$ depends only on the $\sigma$-conjugacy of $\gamma$.)

Finally we record a twisted analogue of Weyl's integration formula for $G(E)$, which is

$$\int_{Z(E)\backslash G(E)} \phi(h)dh = \sum_T |W_T|^{-1}\int\{\int\phi(g^{-\sigma}\gamma g)dg\}\Delta(N\gamma)^2 d\gamma$$

$$= \sum_T |W_T|^{-1}\int F(N\gamma,\phi)\Delta(N\gamma)d\gamma \ .$$

($\gamma$ in $Z(E)T(E)^{1-\sigma}\backslash T(E)$, $g$ in $Z(E)T(F)\backslash G(E)$). The sum is over a set of representatives $T$ for the conjugacy classes of Cartan subgroups in G over F, and $\Delta(N\gamma)$ and $F(N\gamma,\phi)$ are defined by the same formulae used for $h$ in $M_1$. Here $|W_T|$ denotes the number of elements in the Weyl group of $T$ in $G$. We can now deduce:

LEMMA 5. _The character_ $\chi_{I_{P_1}}(\tau,\eta)$ _of_ $I_{P_1}(\tau,\eta)$ _is a function on_ $\sigma \times G(E)$.

_If_ $N\gamma$ _is regular but it does not lie in a split or quadratic torus then_ $\chi_{I_{P_1}}(\tau,\eta)((\sigma,\gamma)) = 0$. _If_ $N\gamma$ _is regular and quadratic then_

$$\chi_{I_{P_1}}(\tau,\eta)((\sigma,\gamma)) = \Delta_1(N\gamma)\chi_{(\tau,\eta)}((\sigma,\gamma))/\Delta(N\gamma).$$

_If_ $N\gamma$ _is regular and split then_

$$\chi_{I_{P_1}}(\tau,\eta)((\sigma,\gamma)) = \sum_w \Delta_1(N(w^{-1}\gamma w))\chi_{(\tau,\eta)}((\sigma,w^{-1}\gamma w))/\Delta(N\gamma) \ ,$$

_where the sum is taken over the set of representatives_ $w$ _for the quotient of the Weyl group of_ $A_0$ _in_ $G$ _by the Weyl group of_ $A_0$ _in_ $M_1$.

In particular, if $\tau$ is the representation of $GL(2,E)$ induced from $P_0 \cap GL(2)$ and the pair $(\eta_1,\eta_2)$ of characters of $E^\times$ with ${}^\sigma\eta_i = \eta_i$, then

$$\chi_{I_{P_1}(\tau,\eta_3)}((\sigma,\gamma)) = \chi_{I_{P_0}(\eta)}((\sigma,\gamma)) = \sum_w \eta(w^{-1}\gamma w)/\Delta(N\gamma) ,$$

for any $\gamma$ such that $N\gamma$ is split and regular, where $\eta = (\eta_1,\eta_2,\eta_3)$ and the sum is taken over the Weyl group of $A_0$ in $G$. The value of this character is $0$ at $\gamma$ for which $N\gamma$ is regular and non-split.

In the proof of Lemma 5 we used a result from the theory of base change for $GL(2)$, namely that the character of any admissible representation on $G \times GL(2,E)$ is a function on the set of $(\sigma,\gamma)$ where $N\gamma$ is conjugate to a regular element in $GL(2,F)$.

1.3.4 Local lifting

Let $H$ be a group defined over $F$, and suppose $\pi$ and $\pi^E$ are admissible irreducible representations of $H(F)$ and $H(E)$ whose characters $\chi_\pi$, $\chi_{\pi^E}$ are functions on the regular set of $H(F)$ and $\sigma \times H(E)$, respectively.

DEFINITION. *The representation* $\pi$ *corresponds to the representation* $\pi^E$ *of* $G(E)$ *if there exists an extension, denoted again by* $\pi^E$, *of* $\pi^E$ *to* $G \times G(E)$, *such that*

$$\chi_{\pi^E}((\sigma,\gamma)) = \chi_\pi(N\gamma)$$

*for all* $\gamma$ *in* $H(E)$ *such that* $N\gamma$ *is conjugate to a regular element in* $H(F)$, *or, equivalently, if* $\operatorname{tr} \pi^E(\phi') = \operatorname{tr} \pi(f)$ *for all* $\phi'$ *on* $\sigma \times G(E)$ *and* $f$ *on* $G(F)$ *such that* $\phi \longrightarrow f$ *in the sense of Lemma 3, where* $\phi(g) = \phi'((\sigma,g))$ *is a function on* $G(E)$.

For example, if $\mu^E$ is a quasi-character of $GL(1,E) = E^\times$ with ${}^\sigma\mu^E = \mu^E$ then there exists a character $\mu$ of $F^\times$ such that $\mu^E(x) = \mu(Nx)$.

Here $\mu$ is defined up to a character of $NE^\times\backslash F^\times$, and each $\mu$ corresponds to $\mu^E$ in the above sense. This definition was used in [12] in the context of GL(2) and will be used here also in the context of $G = GL(3)$.

Lemma 5 has the following application for induced representations which are irreducible (since we have defined the correspondence only for irreducible $\pi$ and $\pi^E$; cf. 6.3.2).

COROLLARY 6. <u>If the representation</u> $(\tau,\eta)$ <u>of</u> $M_1(F)$ <u>corresponds to the representation</u> $(\tau^E,\eta^E)$ <u>of</u> $M_1(E)$ <u>then the representation</u> $I_{P_1}(\tau^E,\eta^E)$ <u>of</u> $G(F)$ <u>corresponds to the representation</u> $I_{P_1}(\tau^E,\eta^E)$ <u>of</u> $G(E)$. <u>In particular, if the representation</u> $\eta$ <u>of</u> $A_0(F)$ <u>corresponds to the representation</u> $\eta^E$ <u>of</u> $A_0(E)$ <u>then the representation</u> $I_{P_0}(\eta)$ <u>of</u> $G(F)$ <u>corresponds to the representation</u> $I_{P_0}(\eta^E)$ <u>of</u> $G(E)$.

The character $\chi_{\sigma(\eta)}$ of the Steinberg representation $\sigma(\eta)$ in $I_{P_0}(\eta)$ $(\eta = (\mu\alpha,\mu,\mu\alpha^{-1}))$ can be expressed as a linear combination of characters of induced representations, namely,

$$\chi_{\sigma(\eta)} = \chi_{I_{P_0}(\eta)} - \chi_{I_{P_2}(\mu\alpha,\pi(\mu,\mu\alpha^{-1}))} - \chi_{I_{P_1}(\pi(\mu\alpha,\mu),\mu\alpha^{-1})} + \chi_\mu,$$

where $\chi_\mu$ denotes the character of the one-dimensional representation $g \longrightarrow \mu(\det g)$ of $G(F)$. If we replace $\alpha$ by the valuation $\alpha_E$ on $E$ and $\eta$ by $\eta^E$ such that $\eta^E(x) = \eta(Nx)$ $(x$ in $E^\times)$ then the above relation remains valid on $(\sigma,\gamma)$ for $\gamma$ in $G(E)$ such that $N\gamma$ is regular; the associated Steinberg representation is denoted by $\sigma^E(\eta^E)$. From Lemma 5 we deduce

COROLLARY 7. *If* $\eta$ *corresponds to* $\eta^E$ *then* $\sigma(\eta)$ *corresponds to* $\sigma^E(\eta^E)$.

As noted above we conform with the conventions used for $GL(2)$ and write $\pi_{P_2}(\eta\alpha^s, \tau \otimes \alpha^{s'})$ for $I_{P_2}(\eta\alpha^s, \tau \otimes \alpha^{s'})$ when it is irreducible. The same notations can be introduced for $P_1$ and for $P_0$. We obtain

COROLLARY 8. *If* $(\eta,\tau)$ *corresponds to* $(\eta^E,\tau^E)$ *then* $\pi = \pi_{P_2}(\eta\alpha^s, \tau \otimes \alpha^{s'})$ (*resp.* $\pi_{P_1}(\tau \otimes \alpha^s, \eta\alpha^{s'})$) *corresponds to* $\pi^E = \pi^E_{P_2}(\eta^E\alpha^s_E, \tau^E \otimes \alpha^{s'}_E)$ (resp. $\pi^E_{P_1}(\tau^E \otimes \alpha^s_E, \eta^E\alpha^{s'}_E)$) *whenever* (i) $\tau$ *is supercuspidal*, (ii) $\tau^E$ *is equal to* $\sigma^E(\mu^E\alpha_E^{1/2}, \mu^E\alpha_E^{-1/2})$ *and either* $\eta^E \neq \mu^E$ *or* $s \neq s' + 3/2$, (iii) $\tau$ *is equal to* $\sigma(\mu\alpha^{1/2}, \mu\alpha^{-1/2})$ *and both* $\eta = \mu$ *and* $s = s' + 3/2$, (iv) $\pi^E$ *is* $\pi^E_{P_0}(\eta^E)(\eta^E = (\eta^E_1, \eta^E_2, \eta^E_3))$ *and none of* $\eta^E_i/\eta^E_j$ $(i > j)$ *is equal to* $\alpha_E$, (v) $\pi^E$ *is* $\pi^E_{P_0}(\eta^E)$ *and* $\pi$ *is* $\pi_{P_0}(\eta)$, $\eta_i/\eta_j = \alpha$ *for a single pair* $(i,j)$ *and* $\eta^E_{i_1}/\eta^E_{j_1} \neq \alpha_E$ *for any pair* $(i_1,j_1)$ *other than* $(i,j)$, (vi) $\pi$ *is* $\pi_{P_0}(\eta)$ *and* $\eta_i/\eta_j = \alpha$ *for all pairs* $i > j$.

Here we assume that $s \geq s'$ and $|\eta^E_i/\eta^E_j| = \alpha_E^{s_1}$ $(s_1 > 0;\ i > j)$. Cases (iv), (v), (vi) deal with $\tau = \pi(\eta_2, \eta_3)$ and $\tau^E = \pi^E(\eta^E_2, \eta^E_3)$. Note that cases (ii), (iii) do not include the case where $s = s' + 3/2$, $\eta^E = \mu^E$ and $\eta \neq \mu$; here $\pi$ does not correspond to $\pi^E$. Also the case $\eta_i/\eta_j \neq \alpha$ and $\eta^E_i/\eta^E_j = \alpha_E$ for any pair $i > j$ is not covered by (iv), (v), (vi) since then $\pi$ does not correspond to $\pi^E$.

For the discussion of the global correspondence we shall extend in 6.3.2 the definition of the local correspondence so as to formally include some of the cases which were excluded here.

1.4.1 Orthogonality relations

Let $\pi_1$ and $\pi_2$ be admissible irreducible square-integrable representations of $G(F)$. Suppose the restrictions of their central characters to $NE^\times$ are equal. If $\pi_1$ is the Steinberg representation then its character exists as a smooth function on the regular set and it is locally integrable on $G(F)$ (see Lemma 5 and Corollary 7; as usual, $h$ in $G(F)$ is regular if $\Delta(h) \neq 0$). Otherwise $\pi_1$ is a supercuspidal representation.

Generalizing a theorem of [8] from the context of $GL(2)$ to the context of any reductive connective p-adic groups over a field of characteristic 0, Harish-Chandra [6] proved that the character of supercuspidal representation exists as a smooth function on the regular set (Theorem 12,p. 60) and it is locally integrable on $G(F)$ (Theorem 16, p. 92). Thus, denoting by $\chi_{\pi_i}$ the character of $\pi_i$ $(i = 1,2)$ and fixing a non-trivial character $\xi$ of $NE^\times\backslash F^\times$ we can state

LEMMA 9. The expression

$$\sum_T (|W_T| |NZ(F)\backslash T(F)|)^{-1} \int_{Z(E)T(E)^{1-\sigma}T(E)} \chi_{\pi_1}(N\gamma)\overline{\chi}_{\pi_2}(N\gamma)\Delta(N\gamma)^2 d\gamma$$

is equal to $1$, $\ell^{-1}$ or $0$ if (i) $\pi_1 \simeq \xi^i \otimes \pi_2$, and $\pi_1 \simeq \xi \otimes \pi_1$ (ii) $\pi_1 \simeq \xi^i \otimes \pi_2$ but $\pi_1 \not\simeq \xi \otimes \pi_1$ (for some $i$ $(0 \leq i < \ell)$),

(iii) $\pi_1 \not\simeq \xi^i \otimes \pi_2$ _for all_ $i$ _(respectively). The sum is taken over a set of representatives of the conjugacy classes of elliptic_ (= _cubic) tori of_ $G(F)$.

_Proof._ For elliptic torus $T$ the set $NT(E)$ is equal to the set of $h$ in $T(F)$ such that $\xi(\det h) = 1$. Hence

$$\ell^{-1} \sum_{i=0}^{\ell-1} \chi_{\xi^i \otimes \pi_j}(h) \qquad (j = 1,2)$$

is equal to $\chi_{\pi_j}(h)$ on $NT(E)$ and to $0$ outside it. Since the bijection $\gamma \longrightarrow N\gamma$ from $Z(E)T(E)^{1-\sigma}\backslash T(E)$ to $NZ(E)\backslash NT(E)$ is measure preserving the lemma follows from the orthogonality relations for the characters of square-integrable representations of $G(F)$.

Let $X$ be the union of the regular elements in $NT(E)$, taken over a set of representatives for the conjugacy classes of Cartan subgroups $T$ of $G$ over $F$. Consider the set $S$ of restrictions to $X$ of the characters of all admissible irreducible representations $\pi$ of $G(F)$. Clearly we obtain the same element in $S$ from $\pi_1$ and $\pi_2$ if either (i) both are square-integrable and $\pi_1 \simeq \xi \otimes \pi_2$ for some $\xi$ as above or (ii) both are of the form described by Corollary 8 and they correspond to the same $\pi^E$.

LEMMA 10. _Any linear relation among the elements of_ $S$ _is generated by the relations_

(i) $\chi_{\sigma(\mu\alpha,\mu,\mu\alpha^{-1})} + \chi_{\pi_{P_1}}(\sigma(\mu\alpha,\mu),\mu\alpha^{-1}) = \chi_{\pi_{P_1}}(\sigma(\mu\alpha,\mu),\xi\mu\alpha^{-1})$

(ii) $\chi_{\sigma(\mu\alpha,\mu,\mu\alpha^{-1})} + \chi_{\pi_{P_2}}(\mu\alpha,\sigma(\mu,\mu\alpha^{-1})) = \chi_{\pi_{P_2}}(\xi\mu\alpha,\sigma(\mu,\mu\alpha^{-1}))$

(iii) $\chi_{\pi(\mu\alpha,\mu,\mu\alpha^{-1})} + \chi_{\pi_{P_2}}(\sigma(\mu\alpha,\mu),\mu\alpha^{-1}) = \chi_{\pi_{P_2}}(\xi\mu\alpha,\pi(\mu,\mu\alpha^{-1}))$

(iv) $\chi_{\pi(\mu\alpha,\mu,\mu\alpha^{-1})} + \chi_{\pi_{P_2}}(\mu\alpha,\sigma(\mu,\mu\alpha^{-1})) = \chi_{\pi_{P_2}}(\pi(\mu\alpha,\mu),\xi\mu\alpha^{-1})$

and the relations induced (from $P_i$ to $G(i = 1,3)$) from the relation

(v) $\chi_{\pi(\mu\alpha,\mu)} + \chi_{\sigma(\mu\alpha,\mu)} = \chi_{\pi(\xi\mu\alpha,\mu)}$

on $GL(2)$.

Proof. Given any such relation we can apply (iii) (or iv)) to obtain a relation without the character of the one-dimensional representation. Applying (i) or (ii) we obtain a relation containing only characters of representations induced from $P_2$ and square-integrable characters. The orthogonality relations of Lemma 9 imply that the relation does not involve characters of square-integrable representations. Hence the relation is a consequence of a relation among characters of representations of $M_2 \cong GL(1) \times GL(2)$. This is generated by (v) (see [12], Lemma 7.14).

1.4.2 Supercuspidals

Let $\pi^E$ be a supercuspidal representation of $G(E)$ with ${}^{\sigma}\pi^E \cong \pi^E$. Then $\pi^E$ can be extended to $G \times G(E)$. We choose a fixed extension and denote it again by $\pi^E$. As mentioned above, the character $\chi_{\pi^E}$ of $\pi^E$

exists as a locally integrable function on the connected subgroup $G(E)$ of $\mathcal{G} \times G(E)$ and it is smooth on the regular set of $G(E)$ ([6], Theorems 12,16).

This result is also valid for the non-connected semi-direct product $\mathcal{G} \times G(E)$. Indeed the assumption that the group be connected is not used in any essential way in [6]. One can also deduce the properties of the character on $\mathcal{G} \times G(E)$ from those already established for $G(E)$ and $G(F)$. Thus we note that the regular set of $\mathcal{G} \times G(E)$ consists of the $\gamma$ in $G(E) = 1 \times G(E)$ with $\Delta_E(\gamma) \neq 0$ and the $(\sigma,\gamma)$ with non-trivial $\sigma$ in $\mathcal{G}$ such that $\Delta_E((\sigma,\gamma)) = \Delta(N\gamma) \neq 0$. The change of variable formula of [6], Lemma 22 (ii) takes the form

$$\int_{N(E)} \phi(\sigma,\gamma n)dn = \delta_E(\gamma)^{1/2}\Delta_E((\sigma,\gamma))\int_{N(E)} \phi((\sigma,n^{-\sigma}\gamma n))dn$$

on the regular $(\sigma,\gamma)$ since $n^{-1}\cdot(\sigma,\gamma)\cdot n = (\sigma,n^{-\sigma}\gamma n)$ ($n$ in $N(E)$).

Theorem 15, p. 86, extends to $\sigma \times G(E)$, since it is valid for $G(F)$, by a double application of the Weyl integration formula:

$$\int_{Z(E)\backslash G(E)} \Delta((\sigma,g))^{-1-\varepsilon}\phi((\sigma,g))dg$$

$$= \sum_T |W_T|^{-1}\int_{Z(E)T(E)^{1-\sigma}\backslash T(E)} \Delta(N\gamma)^{2-1-\varepsilon}\{\int_{Z(E)T(F)\backslash G(E)} \phi((\sigma,g^{-\sigma}\gamma g)dg\}d\gamma$$

$$= \sum_T |W_T|^{-1}\int_{NZ(E)\backslash NT(E)} \Delta(h)^{1-\varepsilon}F(h,\phi)dh$$

$$= \sum_T |W_T|^{-1} \int_{NZ(E)\backslash T(E)} \Delta(h)^{1-\varepsilon} F(h,f)\,dh$$

$$= \int_{NZ(E)\backslash G(F)} \Delta(g)^{-1-\varepsilon} f(g)\,dg \qquad (\text{if } \phi \longrightarrow f).$$

Although it is not necessary we note that Theorem 14 also extends to $G \times G(E)$ since the map $\phi \longrightarrow f$ exists. The remaining estimates for $G \times G(E)$ are obtained from the corresponding ones for $G(F)$ so that the formal proof for Theorem 16 (and 12) establishes the required properties of the character on $G \times G(E)$.

An additional useful property of the character of a supercuspidal representation is that

$$\chi_{\pi^E}((\sigma,\gamma)) = d(\pi^E) \int_{Z(E)\backslash G\times G(E)} (\pi^E(g^{-1}(\sigma,\gamma)g)u,\tilde{u})\,dg$$

on the elliptic set (that is the $(\sigma,\gamma)$ such that either $\sigma = 1$ and $\gamma$ is elliptic or $\sigma \neq 1$ and $N\gamma$ is elliptic). Note that in particular the integral is absolutely convergent. Here $u$ is a vector in the space of $\pi^E$ and $\tilde{u}$ is a vector in the space of the contragredient representation of $\pi^E$ with $(u,\tilde{u}) = 1$. We denote by $d(\pi^E)$ the formal degree of $\pi^E$.

1.4.3 Twisted orthogonality relations

There are a few applications of the Schur orthogonality relations which will be useful for us. Let $\sigma$ be again a fixed generator of $G$. We have

LEMMA 11. *If* $\pi^E$ *is unitary and supercuspidal then*

$$\sum_T |W_T|^{-1} |NZ(F)\backslash T(F)|^{-1} \int |\chi_{\pi^E}((\sigma,\gamma))|^2 \Delta(N\gamma)^2 d\gamma = \ell^{-1},$$

($\gamma$ *in* $Z(E)T(E)^{1-\sigma}\backslash T(E)$) *where the sum is taken over a set of representatives of the conjugacy classes of elliptic* (= *cubic*) *tori of* $G(F)$.

*Proof*. Let $\zeta_1$ be a primitive $\ell^{th}$ root of unity and let $\zeta$ be the character of $G \times G(E)$ which is trivial on $G(E)$ and its value at $\sigma$ is $\zeta_1$. Since the representations $\pi_i^E = \zeta^i \otimes \pi^E$ $(0 \leq i < \ell)$ are inequivalent, the Schur orthogonality relations imply that $\operatorname{tr} \pi_i^E(\phi)$ is equal to 1 if $i = 0$ and to 0 if $1 \leq i < \ell$, where

$$\phi(g) = d(\pi^E)(\pi^E(g)u,\tilde{u}) \qquad ((u,\tilde{u}) = 1).$$

It follows that

$$1 = \sum_{i=0}^{\ell-1} \zeta_1^{-i} \operatorname{tr} \pi_i^E(\phi) = \ell \int_{Z(E)\backslash\sigma\times G(E)} \phi(g)\chi_{\pi^E}(g)dg.$$

Using Weyl's integration formula we can write the above integral as

$$\sum_T |W_T|^{-1} \int \Delta(N\gamma)^2 \chi_{\pi^E}((\sigma,\gamma)) \{ \int \phi((\sigma, g^{-\sigma}\gamma g))dg \} d\gamma.$$

($\gamma$ in $Z(E)T(E)^{1-\sigma}\backslash T(E)$, $g$ in $Z(E)T(F)\backslash G(E)$). The sum is taken over a set of representatives for the conjugacy classes of tori in $G(F)$.

If $T$ is elliptic the inner integral is equal to

$$|Z(F)\backslash T(F)|^{-1} \int_{Z(E)\backslash G(E)} \phi((\sigma, g^{-\sigma}\gamma g))dg$$

$$= (\ell |Z(F)\backslash T(F)|)^{-1} \int_{Z(E)\backslash G\times G(E)} \phi(g^{-1}(\sigma,\gamma)g)dg = |NZ(F)\backslash T(F)|^{-1} \chi_{\pi^E}((\sigma,\gamma)).$$

If $T$ is quadratic or split we may assume it is contained in $M_1$ or $A_0$, change variables on $N_1$ or $N_0$ in the usual way and deduce from [7], Theorem 29, that the inner integral is $0$. The lemma follows.

There is an additional application of the orthogonality relations. Let $\pi_i^E$ $(i = 1,2)$ be irreducible admissible representations of $G(E)$ with equal central character and with ${}^\sigma\pi_i^E \simeq \pi_i^E$. Denote by $\pi_i^E$ also a fixed extension of $\pi_i^E$ to $G \times G(E)$.

LEMMA 12. *If $\pi_2^E$ is supercuspidal, $\pi_1^E$ is unitary and its character exists as a locally integrable function, then*

$$\sum_T |W_T|^{-1} |NZ(F)\backslash T(F)|^{-1} \int \chi_{\pi_1^E}((\sigma,\gamma)) \overline{\chi}_{\pi_2^E}((\sigma,\gamma)) \Delta(N\gamma)^2 d\gamma$$

*($\gamma$ in $Z(E)T(E)^{1-\sigma}\backslash T(E)$) is $0$ if $\pi_1^E$ and $\pi_2^E$ are inequivalent as representations of $G(E)$. The sum is over the same set as in the previous lemma.*

*Proof.* We have $\operatorname{tr}(\zeta^i \otimes \pi_1^E)(\phi) = 0$ $(0 \le i < \ell)$ for

$$\phi(g) = d(\pi_2^E)\overline{(\pi_2^E(g)u,\tilde{u})} \quad (g \text{ in } G \times G(E), (u,\tilde{u}) = 1).$$

Hence

$$0 = \ell^{-1} \sum_{i=0}^{\ell-1} \zeta_1^{-i} \operatorname{tr}(\zeta^i \otimes \pi_1^E)(\phi) = \int_{Z(E)\backslash \sigma \times G(E)} \chi_{\pi_1^E}(g)\phi(g)dg,$$

and the lemma follows from Weyl's integration formula and the usual change of variables, as in the proof of Lemma 11.

1.5.1 Split places

Let $F$ be a field, $E$ a direct sum of $\ell$ copies of $F$, $G$ a linear algebraic group over $F$, $G(F)$ the group of $F$-rational points of $G$ and $G(E) = G(F) \otimes_F E$ is isomorphic to the direct product $G_E(F)$ of $\ell$ copies of $G(F)$. Let $\sigma$ be an automorphism of $G(F)$ of order $\ell$, and embed $G(F)$ in $G_E(F)$ by

$$x \longmapsto (x, \sigma^{\ell-1}(x), \ldots, \sigma(x)).$$

Then $\sigma$ defines an automorphism $\sigma'$ of $G_E(F)$ (denoted by $\sigma$ when the context is clear), given by

$$\sigma' : (x_1, \ldots, x_\ell) \longrightarrow (\sigma x_2, \ldots, \sigma x_\ell, \sigma x_1).$$

If $\mathcal{G}$ is the group generated by $\sigma'$ then $\mathcal{G}$ acts on $G_E(F)$ and the semi-direct product $\mathcal{G} \times G_E(F)$ can be introduced. The norm $N\gamma$ of $\gamma = (\gamma_1, \ldots, \gamma_\ell)$ in $G_E(F)$ is defined to be

$$\begin{aligned} N\gamma &= \sigma'^{\ell-1}(\gamma) \ldots \sigma'(\gamma)\gamma \\ &= (\sigma^{\ell-1}(\gamma_\ell) \ldots \sigma(\gamma_2)\gamma_1, \sigma^{\ell-1}(\gamma_1)\sigma^{\ell-2}(\gamma_\ell) \ldots \sigma(\gamma_3)\gamma_2, \ldots, \\ &\qquad \sigma^{\ell-1}(\gamma_{\ell-1})\sigma^{\ell-2}(\gamma_{\ell-2}) \ldots \sigma(\gamma_1)\gamma_\ell). \end{aligned}$$

This is conjugate in $G_E(F)$ to $h = (\sigma^{\ell-1}(h), \ldots, \sigma(h), h)$ $(h = \sigma^{\ell-1}(\gamma_\ell) \ldots \sigma(\gamma_2)\gamma_1)$.

This situation occurs whenever a place of the global field $F$ splits completely in the global field $E$, and then $\sigma = 1$. We shall deal with arbitrary $\sigma$, and not only with $\sigma = 1$ which is the only case needed here, since there is only a little difference in the proof of the general and special cases, and the general case will be needed elsewhere (e.g. $\sigma(g) = J^t g^{-1} J^{-1}$ for unitary groups).

1.5.2 Matching functions

Suppose $\phi$ is a function on $G_E(F)$ of the form

$$\phi(g) = f_1(g_1)\dots f_\ell(g_\ell) \quad (g = (g_1,\dots,g_\ell)) ,$$

and $f_i$ are smooth compactly supported functions on $G(F)$. Define $f$ on $G(F)$ $(f = f_1 \overset{\sigma}{*} f_2 \overset{\sigma}{*} \dots \overset{\sigma}{*} f_\ell)$ by

$$f(g) = \int f_1(\sigma(m_2^{-1})g)f_2(\sigma(m_3^{-1})m_2)\dots f_{\ell-1}(\sigma(m_\ell^{-1})m_{\ell-1})f_\ell(m_\ell)dm$$

with $m = (m_2,m_3,\dots,m_\ell)$ and $m_i$ in $G(F)$. The following lemma asserts that $\phi \longrightarrow f$.

LEMMA 13. With the above notations we have

$$\int_{G_\gamma^\sigma(E)\backslash G(E)} \phi(\sigma(g^{-1})\gamma g)dg = \int_{G_h(F)\backslash G(F)} f(g^{-1}hg)dg.$$

*Proof.* We change variables on the left side by

$$g \longrightarrow (1,\sigma^{\ell-1}(\gamma_1 h^{-1}),\sigma^{\ell-1}(\gamma_2)\sigma^{\ell-2}(\gamma_1 h^{-1}),\dots,\sigma^{\ell-1}(\gamma_{\ell-1})\sigma^{\ell-2}(\gamma_{\ell-2})\dots\sigma(\gamma_1 h^{-1}))g,$$

and deduce that $\gamma$ can be replaced by $(h,1,\dots,1)$. But if $\gamma = (h,1,\dots,1)$ and $\sigma(g^{-1})\gamma g = \gamma$ then $g = (g_1,\sigma^{\ell-1}(g_1),\dots,\sigma^2(g_1),\sigma(g_1))$ with $g_1$ such that $g_1^{-1}hg_1 = h$ so that $G_\gamma^\sigma(E)$ can be replaced by the subgroup $G_h(F)$ of $G(F)$. The left side becomes

$$\int_{G_h(F)\backslash G(F)\times G(F)\times\ldots\times G(F)} f_1(\sigma(g_2^{-1})hg_1)\ldots f_{\ell-1}(\sigma(g_\ell^{-1})g_{\ell-1})f_\ell(\sigma(g_1^{-1})g_\ell)dg.$$

The new variables $m_i = \sigma^{1-i}(g_1^{-1})g_i$ $(2\leq i\leq \ell)$ afford rewriting this as

$$\int_{(G_h(F)/G(F))\times G(F)\times\ldots\times G(F)} f_1(\sigma(m_2^{-1})g_1^{-1}hg_1)f_2(\sigma(m_3^{-1})m_2)$$

$$\ldots f_{\ell-1}(\sigma(m_\ell^{-1})m_{\ell-1})f_\ell(m_\ell)dm \quad ,$$

which is the right side of the lemma.

1.5.3 Lifting representations

Let $\pi$ be an irreducible admissible representation of $G(F)$ on $V$. The representation $\pi^E = \pi\otimes{}^\sigma\pi\otimes{}^{\sigma^2}\pi\otimes\ldots\otimes{}^{\sigma^{\ell-1}}\pi$ of $G_E(F)$ (where ${}^\sigma\pi(g) = \pi(\sigma(g)))$ extends to $G\times G_E(F)$ by

$$\pi^E(\sigma)\colon v_1\otimes\ldots\otimes v_\ell \longrightarrow v_\ell\otimes v_1 \ldots\otimes v_{\ell-1}.$$

For $f$ as above we choose a basis $\{v_i\}$ for $V$ so that $\operatorname{tr}\pi(f) = \sum \pi_{ii}(f)$ where $\pi_{ij}(f) = (\pi(f)v_i,v_j)$. The matrix of

$$\pi^E(\sigma)\pi^E(\phi) = \pi^E(\sigma)(\pi(f_1)\otimes{}^\sigma\pi(f_2)\otimes\ldots\otimes{}^{\sigma^{\ell-1}}\pi(f_\ell))$$

with respect to the basis $\{v_{i_1}\otimes\ldots\otimes v_{i_\ell}\}$ is

$$\{\pi_{i_1j_2}(f_1)\,{}^\sigma\pi_{i_2j_3}(f_2)\ldots{}^{\sigma^{\ell-1}}\pi_{i_\ell j_1}(f_\ell)\}\ ,$$

so that the trace is

$$\sum \pi_{i_1 i_2}(f_1)^{\sigma}\pi_{i_2 i_3}(f_2)\ldots^{\sigma^{\ell-1}}\pi_{i_\ell i_1}(f_\ell) = \operatorname{tr}\ \pi(f),$$

where $f$ is defined before Lemma 13.

As usual we define the function $\phi'$ on $\sigma \times G_E(F)$ by $\phi'((\sigma,g)) = \phi(g)$ and extend it by $0$ to $G \times G_E(F)$. Denoting by $\pi^E$ also the representation of $G \times G_E(F)$ extended from $G_E(F)$ we have

$$\operatorname{tr}\ \pi^E(\phi') = \operatorname{tr}\ \pi^E(\sigma)\pi^E(\phi)\ ,$$

and we obtain

LEMMA 14. <u>The character</u> $\chi_{\pi^E}$ <u>of</u> $\pi^E$ <u>is a locally integrable function on</u> $\sigma \times G(E)$ <u>which is smooth on the regular set. Moreover</u> $\pi$ <u>corresponds to</u> $\pi^E$, <u>that is,</u> $\chi_{\pi^E}((\sigma,\gamma))$ <u>is equal to</u> $\chi_\pi(h)$ <u>if</u> $h$ <u>in</u> $G(F)$ <u>is conjugate to</u> $N\gamma$ <u>and it is regular</u>.

<u>Proof</u>. Since the character of $\pi$ has the required properties we have

$$\operatorname{tr}\ \pi^E(\phi') = \operatorname{tr}\ \pi(f) = \int_{G_E(F)} f_1(\sigma(m_2^{-1})m_1)f_2(\sigma(m_3^{-1})m_2)$$

$$\ldots f_{\ell-1}(\sigma(m_\ell^{-1})m_{\ell-1})f_\ell(m_\ell)\chi_\pi(m_1)dm\ .$$

Changing variables this becomes

$$\int_{G_E(F)} \phi(m_1,\ldots,m_\ell)\chi_\pi(\sigma^{\ell-1})(m_\ell)\ldots\sigma(m_2)m_1)dm\ ,$$

as required.

## 1.5.4 Weighted integrals

There is one more result concerning $\phi$ and $f$ which we need. For its statement and proof recall (cf. 2.2.1) that

$$v_{M(E)}(g) = \sum_P c_P \left\langle \alpha, \sum_{i=1}^{\ell} H_P(g_i) \right\rangle^q \quad (g = (g_1,\dots,g_\ell) \text{ in } G_E(F))$$

defines a left $M(E)$-invariant function on $G(E)$; here the sum over $P$ is taken over the parabolic subgroups of $G$ whose Levi component is $M$, and the rank $q$ of $P$, the constants $c_P$, and the right $K(F)$-invariant function $H_P$ on $G(F)$ are defined as in [3],§7. Denote by $f^o$ the quotient of the characteristic function of $K(F)$ by the measure of $K(F)$. We shall deal only with $M$ such that $M(F)$ is $\sigma$-stable, that is, $\sigma M(F) = M(F)$, and assume that $H_P$ is $\sigma$-invariant, whence $H_P(\sigma(g)) = H_P(g)$ (g in $G(F)$).

LEMMA 15. *Suppose* $\phi(g) = f_1(g_1)\dots f_\ell(g_\ell)$ *where* $f_i = f^o$ $(1 < i \leq \ell)$ *and* $f_1$ *is* $K(F)$-*biinvariant. If* $\gamma$ *lies in* $M(E)$ *and* $N\gamma$ *is conjugate to a regular element* $h$ *of* $G(F)$ *then*

$$\int_{G_\gamma^\sigma(E)\backslash G(E)} \phi(\sigma(g^{-1})\gamma g) v_{M(E)}(g)dg = \ell^q \int_{G_h(F)\backslash G(F)} f(g^{-1}hg)v_M(g)dg$$

($v_M(g)$ *is defined by the expression for* $v_{M(E)}(g)$ *with* $\sum H_P(g_i)$ *replaced by* $H_P(g)$).

*Proof.* Since $v_{M(E)}(g)$ is left $M(E)$-invariant and $\sigma M(F) = M(F)$, the first change of variables in Lemma 13 shows that the left side above is equal to

$$\int f_1(\sigma(g_2^{-1})hg_1)f_2(\sigma(g_3^{-1})g_2)\ldots f_\ell(\sigma(g_1^{-1})g_\ell)v_{M(E)}(g)dg.$$

The second change of variables there, but with $g = g_1$ and $m_1 = 1$, shows that this is

$$\int f_1(\sigma(m_2^{-1})g^{-1}hg)f_2(\sigma(m_3^{-1})m_2)\ldots f_{\ell-1}(\sigma(m_{\ell-1}^{-1})m_\ell)f_\ell(m_\ell)$$
$$\Big(\sum_P c_P \left\langle\alpha,\textstyle\sum_i H_P(\sigma^{1-i}(g)m_i)\right\rangle^q\Big) .$$

But $f_i = f^o$ $(2 \leq i \leq \ell)$, $H_P$ is right $K(F)$-invariant and $\sigma$-invariant, and $f_1$ is spherical. Hence the expression in brackets is $\ell^q v_M(g)$, and we obtain

$$\ell^q \int f_1(g^{-1}hg)v_M(g)dg .$$

The lemma follows from the fact that $f$ is equal to $f_1$.

Note that at least when $q = 1$ (and $\sigma = 1$) Lemma 15 is valid for any $\ell$-tuple $(f_1,\ldots,f_\ell)$ of $K(F)$-biinvariant functions on $G(F)$ ([12],§9).

1.5.5 Matching operators

We shall need a lemma whose relation to Lemma 14 is analogous to the relation of Lemma 15 to Lemma 13. It will be used in Chapter 5 to obtain a corollary of [12] in the case of $G = GL(2)$ where $\sigma$ is the identity automorphism of $G(F)$, as we now assume. Let $A$ be an operator on the space of the representation $\pi$, and let $A_k$ be the operator on the space of $\pi^E = \pi\otimes\ldots\otimes\pi$, given by $A$ applied to the kth component, namely

$$A_k: v_1\otimes\ldots\otimes v_k\otimes\ldots\otimes v_\ell \longrightarrow v_1\otimes\ldots\otimes Av_k\otimes\ldots\otimes v_\ell .$$

LEMMA 16. *For any* $\phi$ *as above and* $f^k = f_k * f_{k+1} * \ldots * f_\ell * f_1 * \ldots * f_{k-1}$ , *we have*

$$\mathrm{tr}\ A_k \pi^E(\phi') = \mathrm{tr}\ A\ \pi(f^k) ,$$

*and*

$$\mathrm{tr} \sum_{k=1}^{\ell} A_k\ \pi^E(\phi') = \ell\ \mathrm{tr}\ A\ \pi(f)$$

*if* $f^k = f$ *for all* $k$ $(1 \leq k \leq \ell)$.

Only the first claim needs to be proved. The matrix of $A_k\pi^E(\sigma)\pi^E(\phi)$ with respect to $\{v_{i_1}\otimes\ldots\otimes v_{i_\ell}\}$ is

$$\{\pi_{i_1 j_2}(f_1)\pi_{i_2 j_3}(f_2)\ldots A\pi_{i_{k-1} j_k}(f_{k-1})\ldots\pi_{i_\ell j_1}(f_\ell)\} ,$$

and the trace is

$$\sum \pi_{i_k i_{k+1}}(f_k)\ldots\pi_{i_\ell i_1}(f_\ell)\pi_{i_1 i_2}(f_1)\ldots\pi_{i_{k-2} i_{k-1}}(f_{k-2})A\pi_{i_{k-1} i_k}(f_{k-1})$$

$$= \sum \pi_{i_k i_{k-1}}(f_k * \ldots * f_\ell * f_1 * \ldots * f_{k-2})A\pi_{i_{k-1} i_k}(f_{k-1}) = A\pi(f^k),$$

since $A\pi(f_2)\pi(f_1)v = A\pi(f_1 * f_2)v$.

It is the last statement of the lemma which will be used in Chapter 5, with the operator $A = R^{-1}(\eta_v)R'(\eta_v)$ (see [12]) and the function $\phi$ such that (1) $f_1$ is smooth compactly supported (modulo $Z(F)$), hence left-invariant by some small (compact) subgroup $K'$ of $K$, (2) $f_i$ $(2 < i \leq \ell)$ are $0$ outside $Z(F)K'$ and their value on $K'$ is $1/|K'|$ , where $|K'|$ is the volume of $K'$. The identity of Lemma 15 is also valid for such functions $\phi$.

As is already suggested by the last comment, the lemmas are also valid for functions $\phi$ and $f$ which are compactly supported modulo $Z(E)$ and $NZ(E)$ and transform by characters $\omega_E^{-1} = \omega^{-1} \circ N$ and $\omega^{-1}$ on these groups (respectively). This can be deduced from the above lemmas in the usual way ([12], end of §5). It is the latter kind of $\phi$ and $f$ with which we are going to deal below.

## §2. THE TRACE FORMULA

### 2.1.1 Introduction

Let $\omega$ be a unitary character of $Z_E(\mathbb{A}) = Z(F)N_{E/F}Z(\mathbb{A}_E)$ trivial on $Z(F)$. Denote by $L^2(\omega)$ the space of measurable functions $\psi$ on $G(F)\backslash G(\mathbb{A})$ with

$$\psi(zg) = \omega(z)\psi(g) \qquad (z \text{ in } Z_E(\mathbb{A}),\ g \text{ in } G(F)\backslash G(\mathbb{A}))$$

and

$$\int_{Z_E(\mathbb{A})G(F)\backslash G(\mathbb{A})} |\psi(g)|^2 dg < \infty.$$

The regular representation of $G(\mathbb{A})$ on $L^2(\omega)$ is defined by $r(g)\psi(h) = \psi(hg)$. The space $L^2(\omega)$ is the direct sum of the invariant subspace $L_0^2(\omega)$ of square-integrable cusp forms and its orthogonal complement $L_c^2(\omega)$, which is also invariant.

Let $f = \otimes f_v$ be a function on $G(\mathbb{A})$ whose components $f_v$ are smooth (that is locally constant in the non-archimedean case and highly differentiable and $K_v$-finite for infinite $v$), compactly supported modulo $N_{E_v/F_v}Z(E_v)$ and transform on this group by $\omega_v^{-1}$. For almost all non-archimedean $v$ we have $f_v = f_v^0$. Here $f_v^0$ is the function which obtains the value $0$ unless $g_v = z_v k_v$ ($z_v$ in $NZ(E_v)$, $k$ in $K_v$) when it is the quotient of $\omega_v(z_v)^{-1}$ by the measure of $K_v$.

The restriction of the operator $r(f)$, defined by

$$r(f)\psi(h) = \int_{NZ(\mathbb{A}_E)\backslash G(\mathbb{A})} \psi(hg)f(g)dg,$$

to $L_0^2(\omega)$, is of trace class. Arthur [1] obtained a formula for the trace of the restriction, generalizing works of Selberg, Duflo-Labesse and Jacquet-Langlands, and this trace formula is the difference between two absolutely convergent sums:

$$\operatorname{tr} r(f) = \sum J_{\mathcal{O}}(f) - \sum J_\chi(f). \tag{1}$$

Two elements in $G$ are said to be equivalent if their semi-simple parts are conjugate. The first sum is over all equivalence classes in $G(F)$ modulo $NZ(E)$.

Explicit expressions for $J_{\mathcal{O}}$ are available [1] if $\mathcal{O}$ is a regular class (that is, it contains--and hence consists of--regular--hence semi-simple--elements, so that it is a conjugacy class). For our applications it seems necessary to obtain explicit expressions also for the singular classes. This is one of the purposes of this chapter (see Lemmas 3,4 and sections 2.4,2.5).

The trace formula was expressed by Arthur [3] also in terms of absolutely convergent sums of invariant distributions (to be described below),

$$\operatorname{tr} r(f) = \sum I_{\mathcal{O}}(f) - \sum I_{\chi}(f), \tag{2}$$

by incorporating the "non-invariant" part of the integral distributions $J_{\chi}(f)$ (which are certain integrals over a dual space, of characters) with the $J_{\mathcal{O}}(f)$ to obtain the invariant distributions $I_{\mathcal{O}}(f)$. The trace formula for $GL(2)$ had been put in invariant form by Langlands in order to obtain the applications of [12]. But in [12] a "dual" approach was used, and applying the Poisson summation formula some "non-invariant" parts of the $J_{\mathcal{O}}(f)$ were expressed as integrals (of Fourier transforms of orbital integrals of $f$), and incorporated with the $J_{\chi}(f)$ to form the invariant new distributions.

At first glance it appears that the singular behaviour of $J_{\mathcal{O}}(f)$ (as a function of $\mathcal{O}$) will prevent any further application (as in [12]) of the summation formula, and that the fact that it was at all possible in the case of $GL(2)$ was only accidental. However, by means of a simple, but essential, trick, the $J_{\mathcal{O}}(f)$ can be "corrected" so that they become if not smooth at least regular functions.

Another purpose of this chapter is to study the asymptotic behaviour at the singular set of the "corrected" $J_{\mathcal{O}}(f)$ (see 2.7.1, 2.7.2). A simple calculation (Lemma 8) will show that the summation formula can be applied to such non-smooth functions. It will be clear (Lemma 5) that the limits of the $J_{\mathcal{O}}(f)$ (for split regular $\mathcal{O}$) at the singular set are a "main" part of the $J_{\mathcal{O}}(f)$ with singular $\mathcal{O}$.

It might have been better for obtaining our applications to follow the latter approach (of [12]) in deducing an invariant trace formula from (1). Until this is done, since [3] is already available it will be simpler to apply the summation formula to the $I_{\mathcal{O}}(f)$ of (2). In chapter 5 we shall show that the expression that we thus obtain, together with its twisted analogue (to be given in chapter 3) affords a deduction of a certain equality of traces, from which our applications will be derived in chapter 6.

## 2.1.2 Measures

Since it is important for us to express the global distributions on the right of (1) and (2) in terms of local distributions we shall use Tamagawa measures locally and globally. Thus we fix a non-trivial additive character $c = \otimes c_v$ of $\mathbb{A}$ which is trivial on $F$ and denote by $dx_v$ the Haar measure on $F_v$ self-dual with respect to $c_v$. On $F_v^\times$ we take the Haar measure $d^\times x_v = L(1,1_v)dx_v/|x_v|$. The Tamagawa measure on $\mathbb{A}^\times$ is given by

$$d^\times x = (\lambda_{-1})^{-1} \otimes d^\times x_v, \qquad \lambda_{-1} = \lim(s - 1)L(s,1_F) \quad (s \longrightarrow 1).$$

The measure $dt/t$ on $\mathbb{R}_+^\times$ and the isomorphism $x \longrightarrow |x|$ from $F^0(\mathbb{A})\backslash\mathbb{A}^\times$ to $\mathbb{R}_+^\times$ give a measure on the group $F^0(\mathbb{A})$ (of elements of volume 1 in $\mathbb{A}^\times$) which assigns the measure 1 to the space $F^\times\backslash F^0(\mathbb{A})$.

On $\mathbb{A}$ the Haar measure $dx = \otimes dx_v$ is self-dual, and this gives a measure on $N_i$ with $dn_i = \otimes dn_{iv}$ $(1 \le i \le 3)$. Similarly we obtain measures on $A_i$; for example on $Z(\mathbb{A})\backslash A_0(\mathbb{A})$ we have the relation $da = (\lambda_{-1})^{-2} \otimes da_v$.

We choose an invariant form $\omega$ on $Z\backslash G$ of maximal degree defined over F, and obtain Haar measures $dg = |\omega(g)|$ and $dg_v = |\omega_v(g_v)|$ on $Z(\mathbb{A})\backslash G(\mathbb{A})$ and $Z(F_v)\backslash G(F_v)$ with $dg = \otimes dg_v$. Similarly we choose measures on $A_i\backslash M_i$ which satisfy the same relation. The equation

$$\int_{Z\backslash G} h(g)dg = \int_{Z\backslash A_i}\int_{A_i\backslash M_i}\int_{N_i}\int_{M_i\cap K\backslash K} h(amnk)\,da\,dm\,dn\,dk$$

determines a single Haar measure on $M_i \cap K\backslash K$ globally and locally, and they are related by $dk = (\lambda_{-1})^j \otimes dk_v$ where $j = 1$ if $i = 1,2$ and $j = 2$ if $i = 0$ (since $da = (\lambda_{-1})^{-j} \otimes da_v$). In particular we have $dk = \lambda_{-1} \otimes dk_v$ on $M_i \cap K$ $(i = 1,2)$. On discrete groups we take the measure which assigns 1 to each point, on quotient spaces we take quotient measures (unless otherwise stated), and on the dual $D(H)$ of a locally compact group $H$ we take the measure dual to that on H.

2.1.3. The map H .

Following the notations (and definitions) of [1,2,3] we let $X(M)$ be the group of characters of $M/Z$ defined over F, and put $A_i = \mathrm{Hom}(X(M_i)_F, \mathbb{R})$; its dual space is $A_i^* = X(M_i)_F \otimes \mathbb{R}$. Signify by $\Delta_i$ the set of simple roots of $(P_i, A_i)$. We may assume that $\Delta_0$ consists of the roots $\alpha_1, \alpha_2$, which are the characters of $A_0$ given by

$$\alpha_1(\mathrm{diag}(a,b,c)) = a/b, \qquad \alpha_2(\mathrm{diag}(a,b,c)) = b/c.$$

Then $\Delta_1 = \{\alpha_2\}$, $\Delta_2 = \{\alpha_1\}$. The dual basis $\hat{\Delta}_0$ of $\Delta_0$ consists of weights $\mu_1$, $\mu_2$.

Then $\hat{\Delta}_1 = \{\mu_2\}$, $\hat{\Delta}_2 = \{\mu_1\}$, and $\hat{\Delta}_i$ is a basis for the vector space $A_i$, and $\Delta_i$ is a basis for $A_i^*$. Write $\hat{\tau}_i = \hat{\tau}_{P_i}$ for the characteristic function of the $H$ in $A_0$ with $\langle\mu,H\rangle > 0$ for all $\mu$ in $\hat{\Delta}_i$. We write $T$ for an element in $\mathbb{R}^3$ such that $t_i = \langle T,\mu_i\rangle / \langle\mu_i,\mu_i\rangle$ $(i = 1,2)$ are sufficiently large.

The two dimensional rector space $A_0$ is ismorphic to the space of $(x,y,z)$ in $\mathbb{R}^3$ with $x + y + z = 0$, and then $\alpha_1 = (1,-1,0)$, $\alpha_2 = (0,1,-1)$, and $\mu_1 = (\frac{2}{3}, -\frac{1}{3}, -\frac{1}{3})$, $\mu_2 = (\frac{1}{3}, \frac{1}{3}, -\frac{2}{3})$. The Weyl group of $A_0$ in $G$ is denoted here by $W$. Its elements will be denoted by $w_{id}$(identity), $w_{\alpha_1}$, $w_{\alpha_2}$, $w_{\alpha_1\alpha_2\alpha_1} = w_{\alpha_1}w_{\alpha_2}w_{\alpha_1}$ (reflections with respect to $\alpha_1$, $\alpha_2$, $\alpha_1 + \alpha_2$), and (the rotations) $w_{\alpha_1\alpha_2} = w_{\alpha_1}w_{\alpha_2}$, $w_{\alpha_2\alpha_1} = w_{\alpha_2}w_{\alpha_1}$. The action of $W$ on $A_0$ is described by: $w_{\alpha_1}(x,y,z) = (y,x,z)$, $w_{\alpha_2}(x,y,z) = (x,z,y)$, as these reflections generate $W$. We fix a set of representatives for $W$ in $K(F)$, denoted again by $w_{\alpha_1}$, ... .

There is a map $H_i$ from $M_i$ to $A_i$ (globally and locally) defined by

$$\langle H_i(m),\chi\rangle = \log|\chi(m)| ( = \textstyle\sum_v \log|\chi(m_v)|_v) \qquad (\chi \text{ in } A_i^*).$$

For example for $a_0 = \mathrm{diag}(a,b,c)$ we have

$$H_0(a_0) = \langle H_0(a_0), \alpha_1\rangle \mu_1 + \langle H_0(a_0), \alpha_2\rangle \mu_2$$

$$= \log|a/b|\cdot\mu_1 + \log|b/c|\cdot\mu_2 = (\log|a|, \log|b|, \log|c|) - \frac{1}{3}\log|abc|(1,1,1).$$

and

$$H_1(\mathrm{diag}((\begin{smallmatrix} a & b \\ c & d \end{smallmatrix}), e)) = \frac{1}{2}\log|ad - bc/e^2|\cdot\mu_2.$$

It is extended to $G$ as a right K-invariant and left $N_i$-invariant function by means of the decomposition $G = N_iM_iK$.

If $g = (g_v)$ is an element of $G(\mathbb{A})$ then $H(g) = \sum_v H_v(g_v)$ where $H_v$ is the local analogue of the global function $H$. If $g = (g_{ij})$ is an element of $GL(m,F_v)$ (rows are denoted by $i$, columns by $j$), we can write $g = nak$ with $n$ in the unipotent radical of the standard Borel subgroup, $k$ in the standard maximal compact subgroup and diagonal $a = \mathrm{diag}(a_m, \ldots, a_2, a_1)$. Since the function $H_v$ of $GL(m,F_v)$ ([3]) depends only on $a$ it is useful to note that $|a_1 a_2, \ldots, a_i|_v (1 \leq i \leq m)$ is equal to the norm of the determinants of all $i \times i$ matrices formed from the lines of the $i$ bottom rows of $g$. Thus

$$|a_1|_v = \|(g_{m1}, g_{m2}, \ldots, g_{mn})\|_v, \quad |a_1 a_2|_v = \left\|\left(\det\begin{pmatrix} g_{m-1,j} & g_{m-1,k} \\ g_{m,j} & g_{m,k}\end{pmatrix}; j \neq k\right)\right\|_v, \ldots$$

where $\|(a,b, \ldots)\|_v$ denotes the maximum of $|a|_v, |b|_v, \ldots$ in the non-archimedean case, and the square root of the sum of the squares of $|a|_v, |b|_v, \ldots$ in the archimedean case.

### 2.2.1 The distribution $J_o$

For any equivalence class $o$, in order to define $J_o(f)$ consider the expression

$$\int_{G_F Z_{\mathbb{A}} \backslash G_{\mathbb{A}}} \sum_P (-1)^{|A/Z|} \sum_\delta \hat{\tau}_P(H_0(\delta g) - T) \sum_\gamma \sum_\eta \int_{N_{\gamma_s}(\mathbb{A})} f(g^{-1}\delta^{-1}\eta^{-1}\gamma n \eta\gamma g)\, dn\, dg, \qquad (3)$$

where the sum over $P$ is taken over $G$, $P_1$, $P_2$, $P_0$; $|A/Z|$ denotes the rank of $A = A_P$, $\hat{\tau}_G$ is identically 1. The inner sums are over $\delta$ in $P(F)\backslash G(F)$, $\gamma$ in $M_F \cap Q$ and $\eta$ in $N_{\gamma_s}(F)\backslash N(F)$; $\gamma_s$ denotes the semi-simple part of $\gamma$. It is a polynomial in

$$t_i = \frac{<T,\mu_i>}{<\mu_i,\mu_i>}$$

([3]), Prop. 2.3), and its constant term is equal to $J_o(f)$.

An alternative, more explicit definition for $J_o(f)$ can be given by means of the right K-invariant, left $M_i$-invariant function $v_i(g) = v_{M_i}(g)$. It is given by

$$v_1(g) = -(<\mu_2,H_1(g)> + <\mu_1,H_2(w_{\alpha_1\alpha_2}g)>)/<\mu_2,\mu_2> = \frac{3}{2}\log\|(1,n_2,n_3)\|$$

for $i = 1$. The first equality is obtained by directly calculating the integral over $a$ in $Z(\mathbb{A})A_1(F)\backslash A_1(\mathbb{A})$ of (8.5) from [1] after suppressing $\delta$ and replacing $x$ by $ax$. The second is easily verified for $g = m_1n_1k$ (according to the decomposition $M_1N_1K$ of $G$), with

$$n_i = \begin{pmatrix} 1 & 0 & n_2 \\ & 1 & n_3 \\ & & 1 \end{pmatrix},$$

using the final comment in 2.1.3.

The function $v_G$ is identically 1, and $v_0(g)$ is given by

$$v_0(g) = \frac{3}{2}\sum_w \frac{<\alpha, w^{-1}H_0(wg)>^2}{\Pi_i<\alpha,w^{-1}\alpha_i>} \qquad (i = 1,2)$$

independently of $\alpha = (x,y,z)$ in $\mathbb{R}^3$. We shall calculate this function explicitly in 2.3.2.

Let $h$ be a semi-simple element in $o$, and let $M$ denote the minimal group among $G$, $M_1$, $M_o$ (note that $M_1$ and $M_2$ are conjugate) which intersect $o$ non-trivially. If $h$ is regular then ([1,3]) $J_o(f)$ is given by

$$J_{o}(f) = |Z(\mathbb{A})G_h(F)A(\mathbb{A})\backslash G_h(\mathbb{A})| \int_{G_h(\mathbb{A})\backslash G(\mathbb{A})} f(g^{-1}hg)\ v_M(g)dg.$$

The element $h$ and its class $o$ are named cubic (or elliptic), quadratic or split if $M$ is $G$, $M_1$ or $M_0$ (respectively). The $J_o(f)$ are weighted orbital integrals which are actually orbital integrals if $h$ is elliptic.

2.2.2 Elliptic terms

For the comparison of the trace formulae in chapter 5 we shall now describe the distributions $J_o(f)$ in our context; in particular we shall use $Z_E(\mathbb{A})$ instead of $Z(\mathbb{A})$. We shall do this by rewriting the first sum in (2) as a sum of several subsums; see Lemmas 1, 2, 5 below. The first of these is the following.

LEMMA 1. The subsum over the elliptic $o$ of the (invariant) distributions $J(h,f) = J_o(f)$ ($h$ in $o$) in (2) is equal to the sum over all conjugacy classes of elliptic elements $h$ in $G(F)$ modulo $NZ(E)$ of

$$\varepsilon(h)|Z_E(\mathbb{A})G_h(F)\backslash G_h(\mathbb{A})| \int_{G_h(\mathbb{A})\backslash G(\mathbb{A})} f(g^{-1}hg)dg.$$

The value of $\varepsilon(h)$ is $1/3$ or $1$ depending on whether the equation $g^{-1}hg = zh$ can or cannot be solved with $g$ in $G(F)$ and $z \neq 1$ in $NZ(E)$. Note that since $E/F$ is cyclic $NZ(E)$ is equal to $Z(F) \cap NZ(\mathbb{A}_E)$.

This sum together with (3) of Lemma 4 below, can be considered to be the main part of the trace formula, and it is equal to the corresponding "main part" of the twisted trace formula. The Poisson summation formula will be applied only to the remaining subsums which are sometimes called "cusps".

### 2.2.3 *Quadratic terms*

LEMMA 2. *The subsum over the quadratic* $o$ *of the (non-invariant) distributions* $J(h,f) = J_o(f)$ *in* (2) *is given by*

$$\lambda_{-1} \sum_h \sum_a |Z_E(\mathbb{A})G_h(F)A_1(\mathbb{A})\backslash G_h(\mathbb{A})| \sum_v J(ah,f_v)\Pi_{w\neq v}F(ah,f_w).$$

*For each* $h$ *the functions* $J(ah,f_v)$ *and* $F(ah,f_v)$ *are smooth compactly supported (modulo* $NA(E_v)$*) functions of* $a$ *in* $NA_1(E_v)$. *If, for some place* $v_0$, *all components* $f_v (v \neq v_0)$ *are fixed, then the sum over* $v$ *is taken over a fixed finite set depending only on* $h$ *modulo* $NA_1(E)$ *and on the* $f_v (v \neq v_0)$.

The first sum is taken over a set of representatives $h$ in $M_1(F)$ for the conjugacy classes of quadratic elements in $G(F)$, modulo $NA_1(E)$; the second is over a set of representatives $a$ for $NA_1(E)$ modulo $NZ(E)$.

Note that if $g^{-1}hg = zh$ with some $g$ in $G(F)$, $h$ in $M_1(F)$, and $z$ in $Z(F)$ then $g$ must lie in $M_1(F)$, $g^{-1}hg$ and $h$ must share the non-quadratic eigenvalue and hence $z$ must be 1.

The distribution on $F(h,f_w)$ is as usual the orbital integral of $f_w$ multiplied by $\Delta_w(h)$. It can be termed the normalized orbital integral. The local distribution $J(h,f_v) = J_o(f_v)$ is defined by the product of $\Delta_v(h)$ and the local analogue of the global integral defining $J(h,f)$.

To obtain our expression we used the product formula on $F$ which implies that $\Pi_v\Delta_v(h) = 1$ for any regular $h$ in $G(F)$, and the relation $dk = \lambda_{-1} \otimes dk_v$ between the global and local measures which we chose on $M_1 \cap K\backslash K$. To study $J(ah,f_v)$ as a function of $a$ we drop $v$ (for brevity) and rewrite $J(h,f_v)$ in the form (see 2.2.1)

$$\Delta(h\int_{M_{1h}\backslash M_1}\int_{N_1}\int_{M_1\cap K\backslash K} f(k^{-1}n^{-1}m^{-1}hmnk)\,\tfrac{3}{2}\log\|(1,n_2,\ n_3)\|$$

$$=\tfrac{3}{2}|(h_1-h_2)/e|\iiint f(k^{-1}m^{-1}hmuk)\,\log\|(1,u_2,u_3)\| \qquad (*)$$

where $w = w_{\alpha_1\alpha_2}$, $h_1$ and $h_2$ denote the quadratic eigenvalues of $h$ and $e$ the one which lies in $F^\times$, we put $x = h_1h_2$ and $z = (h_1-e)(h_2-e)/x$, and

$$\begin{pmatrix} u_2 \\ u_3 \end{pmatrix} = z^{-1}\begin{pmatrix} 1-de/x & -be/x \\ -ce/x & 1-ae/x \end{pmatrix}\begin{pmatrix} n_2 \\ n_3 \end{pmatrix}$$

$$m^{-1}hm = \begin{pmatrix} a & b & 0 \\ c & d & 0 \\ & & e \end{pmatrix} \quad \text{and} \quad u = \begin{pmatrix} 1 & 0 & u_2 \\ & 1 & u_3 \\ & & 1 \end{pmatrix}$$

It follows that for every $h$ the function $J(ah,f_v)$ is a smooth compactly supported (modulo $NZ(E_v)$) function of $a$ in $NA_1(E_v)$. The same change of variables shows that $F(h,f_w)$ has the same property.

For the last assertion of the lemma fix $h$ and $f_v(v \neq v_0)$. If $f(k^{-1}m^{-1}ahmnk)$ is non-zero then the fact that $f_w$ are compactly supported modulo $NA(E_w)$ implies that there are $C_w \geq 1$ $(w \neq v_0)$ with $C_w = 1$ whenever $f_w = f_w^0$, so that the quotient of the eigenvalues of $a$ lies between $C_w$ and $C_w^{-1}$ in the w-valuation. The product formula on $F^\times$ implies this for the remaining $v_0$ with $C_{v_0} = \Pi C_w$ $(w\neq v_0)$. Since a discrete subset of a compact set is finite, the sum over $a$ is finite. For each such $a$ our claim is that $J(ah,f_v) \neq 0$ only for a finite set of $v$. Indeed, for almost all $v$ we have $|z|_v = |x/e^2|_v = 1$. If $f_v = f_v^0$ then $a/e$, $b/e$, $c/e$, $d/e$, $n_2$, $n_3$ are bounded by $1$ in the v-valuation. Hence the weight factor in (*), as well as $J(ah,f_v)$, is 0, as claimed.

## 2.3.1 Correction for $GL(2)$

It remains to deal with the split classes $\mathcal{O}$. Since we would like to apply the summation formula to the sum of $J_{\mathcal{O}}(f)$ for such $\mathcal{O}$, we need an analogue of Lemma 2. Unfortunately the local function $J_{\mathcal{O}}(f_v)$ is not even regular as a function of $h$ in the intersection of the split $\mathcal{O}$ and $A_0(F_v)$. To overcome this difficulty we shall define "corrected" weighted orbital integrals $J(h,f)$ locally and globally for split regular $h$ (or $\mathcal{O}$) by replacing the weight factor $v_0(g)$ in $J_{\mathcal{O}}(f)$ with a "corrected" weight factor, locally and globally. Globally we shall have $\sum J(h,f) = J_{\mathcal{O}}(f)$ ($h$ in $\mathcal{O} \cap A_0(F)$); locally the new functions $J(h,f_v)$ will be amenable to the application of the summation formula, in particular continuous, although non-smooth.

For reasons which will be clear only after the definition of the invariant corrected distributions $I(h,f)$ is given, we have to discuss the correction of weighted integrals on $M_1$ as well. As $A_1 \backslash M_1$ is isomorphic to $PGL(2)$ it will suffice to discuss this for $GL(2)$. The discussion in the context of $GL(2)$ is easier and can serve as an introduction to that for $GL(3)$.

The usual notations for $GL(2)$ will now be used; in particular $H = \sum H_v$ is the left $K$-, right $N$-invariant function with $H\begin{pmatrix} a & 0 \\ 0 & b \end{pmatrix} = \log|a/b|\alpha_1$, $\alpha_1 = (1,-1)$. For $g = bk$ with $k = (k_v)$, $k_v$ in the usual compact subgroup $K_v$ of $GL(2,F_v)$, and $b = \begin{pmatrix} a & 0 \\ 0 & b \end{pmatrix}\begin{pmatrix} 1 & n \\ 0 & 1 \end{pmatrix}$, the weight factor is

$$v(g) = \sum_w \langle -w\alpha, H(wg)\rangle / \langle \alpha, w^{-1}\alpha_1\rangle \qquad (\text{any } \alpha = (x,y) \text{ in } \mathbb{R}^2,\ x \neq y)$$

$$= v(b) = \sum_w \langle -w\alpha, \sum_v H_v\left(w\begin{pmatrix} a_v & a_v n_v \\ 0 & b_v \end{pmatrix}\right)\rangle / \langle \alpha, w^{-1}\alpha_1\rangle$$

$$= -\sum_v \left[\langle \alpha, \log|a_v/b_v|\alpha_1\rangle / \langle \alpha,\alpha_1\rangle + \langle -\alpha\ ,\ H_v\begin{pmatrix} \|b_v/(1,n_v)\|_v & \\ & \|a_v(1,n_v)\|_v \end{pmatrix}\rangle / \langle \alpha, -\alpha_1\rangle\right]$$

$$= 2\sum_v \log\|(1,n_v)\|_v \qquad \left(w = \mathrm{id}, \begin{pmatrix} 0 & 1 \\ 1 & 0 \end{pmatrix}\right).$$

If $o$ is the equivalence (= conjugacy) class of a regular $h = \begin{pmatrix} a & \\ & b \end{pmatrix}$ in $GL(2,F)$, then the weighted orbital integral $J_{1o}(f)$ of $f$ at $h$ is given by

$$J_{1o}(f) = \ell \int_{G_h(\mathbb{A})\backslash G(\mathbb{A})} f(g^{-1}hg)v(g)dg$$

$$= 2\ \ell \int f^K(n^{-1}hn) \sum_v \log \|(1,n_v)\|_v dn \qquad (n = (n_v) \text{ in } N(\mathbb{A}),\ n_v = \begin{pmatrix} 1 & n_v \\ 0 & 1 \end{pmatrix} \text{ in } N(F_v)),$$

where $f^K(g) = \int f(k^{-1}gk)dk(k \text{ in } K)$; $\ell$ is the index of $Z_E(\mathbb{A})$ in $Z(\mathbb{A})$. Since $h^{-1}n^{-1}hn$ is equal to $\begin{pmatrix} 1 & n(1-b/a) \\ 0 & 1 \end{pmatrix}$, replacing $n$ by $n(1-b/a)^{-1}$ we see that

$$J_{1o}(f) = 2\ell \int f^K(hn) \sum_v \log \|(1,n_v/(1-b/a))\|_v dn.$$

If now $b/a \longrightarrow 1$ in $\mathbb{A}$, the weight factor becomes infinite for any $n = (n_v)$ in a compact set, and $J_{1o}(f)$ has a singularity as $h$ approaches the singular set in $A(\mathbb{A})$.

Recall that for the trace formula the values of $J_{1o}(f)$ which matter are only those at the F-rational classes $o$ (those which contain an F-rational $h$). The last expression for $J_{1o}(f)$ suggests a way of rewriting it (as a function of $h$ in $A(\mathbb{A})$) without changing its value at the $h$ in $A(F)$, so that its singularities are removed. We simply define the corrected weighted orbital integral to be

$$J_1(h,f) = \ell \int f^K(hn) \sum_v \log \|(1-b/a,n_v)\|_v dn,$$

and note that since $1-b/a$ lies in $F^\times$, the product formula on $F$ shows that

$$\sum J_1(h,f) = 2J_1(h,f) = J_{1o}(f) \qquad (h \text{ in } A(F) \cap o)$$

for any regular $o$.

Since comparison of trace formulae is done in terms of local distributions, we shall write $J_1(h,f)$ in terms of the local distributions $F(h,f_v)$. $J_1(h,f_v)$ is defined by the product of $\Delta_v(h) = |a-b|_v/|ab|_v^{\frac{1}{2}}$ and the local analogue of the integral which defines $J_1(h,f)$ globally. $F(h,f_v)$ is defined by the same expression with the factor $\log\|(1-b/a,n_v)\|_v$ omitted. Indeed the sum over $\mathcal{O}$ which contain a regular split element in $GL(2,F)$ of $J_{1\mathcal{O}}(f)$, which appears in the trace formula for $GL(2)$, is equal to

$$\sum_h J_1(h,f) = \ell\lambda_{-1} \sum_h \sum_v J_1(h,f_v) \prod_{w\neq v} F(h,f_w). \qquad (h \text{ regular in } A(F)).$$

The factor $\lambda_{-1}$ appears since $d^\times a = (\lambda_{-1})^{-1} \otimes d^\times a_v$.

We still have to check whether the function $J_1(h,f)$ of $h$ is continuous, and moreover ameanable for the application of the summation formula. Clearly it is divergent unless the sum over $v$ in its definition is finite. Those familiar with [12] will note that we have to consider simultaneously all functions $f$ such that for a fixed place $v_0$ all components $f_v$ of $f$ are fixed for $v \neq v_0$, but $f_{v_0}$ may vary.

We claim that for any $f$ with fixed $f_v(v\neq v_0)$ the sum over $v$ is taken over any set which contains a fixed finite set which depends only on $f_v(v\neq v_0)$, which includes all $f_v$ with $f_v \neq f_v^0$;$f_v^0$ is the function which vanishes outside $Z(F_v)K_v$, whose value on $K_v$ is $|K_v|^{-1}$. To prove the claim we note that there is a finite set of $h$ for which there exists $n$ in $N(\mathbb{A})$ such that $f(hn)$ is non-zero. Indeed since $f_v$ are compactly supported there are $C_v \geq 1(v\neq v_0)$, $C_v = 1$ for $f_v = f_v^0$, such that $|a/b|_v$ lies between $C_v$ and $C_v^{-1}(v\neq v_0)$. This holds also for $v = v_0$ with $C_{v_0} = \prod C_v(v\neq v_0)$ by the product formula on $F^\times$. Since a discrete subgroup of a compact group is finite, we indeed have only a finite number of $h$.

For each such $h$ we have $|1-b/a|_v = 1$ for almost all $v$, and $J_1(h,f_v^0) = 0$ for such $v$, whence the assertion concerning $v$. It follows that $J_1(h,f)$, in whose definition the sum over $v$ extends over a fixed finite set, extends to a continuous function on $A_0(\mathbb{A})$.

The limit of $J_1(h,f)$ at the singular set is the weighted orbital integral in the following well known expression for $J_{1o}(f)$, with $o$ containing a singular (scalar) $h$:

$$J_{1o}(f) = \ell\lambda_{-1} \sum_v \int f_v^K(hn) \log|n|_v dn \prod_{w \neq v} \int f_w^K(hn)dn$$

$$+ \ell\lambda_{-1}(\lambda_0/\lambda_{-1} - \sum_v L_v'(1)/L_v(1)) \sum_v \int f_v^K(hn)dn + \ell|G(F)Z(\mathbb{A})\backslash G(\mathbb{A})|f(h).$$

The two sums over $v$ extend over the places of $F$ where $f_v \neq f_v^0$. Since

$$\int f_v^K(hn) \log|n|_v dn = L_v'(1)/L_v(1) \int f_v^K(hn)dn$$

for $f_v = f_v^0$, where $L_v(s) = L(s,1_v)$ is the local factor at $v$ of the Hasse-Tate L-function of $F$, the sums can be taken over any set (finite set!) containing all $v$ with $f_v = f_v^0$.

The asymptotic behavior of $J_1(h,f)$ at the singular set is easily found. As it is gentler than that of $J(h,f)$ (see below) on $GL(3)$, we shall discuss below only the case of $GL(3)$.

The function $J_1(h,f)$ is now ready for an application of the summation formula on $A(F)$, $A(\mathbb{A})$. An alternative method was used in [12], where $\frac{1}{2}J_{1o}(f)$ was expressed as a sum of functions $A_{2v}$, $A_{3v}$ of $h$. In addition to being generalizable to higher rank groups, the correction method offered here eliminates (in the case of $GL(2)$) the need to introduce $A_{2v}$ and the calculations of [12], §9.

Finally we note that an alternative definition of $J_1(h,f)$ is given by replacing $H(wg)$ in $v(g)$ of $J_{1o}(f)$ by $H(wg) - \tilde{H}(w\rho)$, where $\rho$ is defined by the equation $h^{-1}\rho^{-1}h\rho = u$, and where $u$ is the regular unipotent element $\begin{pmatrix}1&1\\0&1\end{pmatrix}$ (which has 1's in the diagonal and 1's in the row above it, and o's everywhere else). $\tilde{H}_v(w\rho)$ is defined by the function which describes $H_v(w\rho)$ for $|1-b/a|_v < 1$ in the non-archimedean case, namely $0$ if $w = id$ and $-\log|1-b/a|_v\alpha_1$ otherwise, and $\tilde{H}(w\rho) = \sum_v \tilde{H}_v(w\rho)$ globally. This introduction of $\tilde{H}(w\rho)$ imitates the way in which the singularity of $H(wg)$ is created but with a fixed $u$ replacing the variable $n$. For F-rational $h$ the vector $\tilde{H}(w\rho)$ vanishes by the product formula on F. Although the possibility of correction was first noticed in this last context, we preferred to elaborate on the previous more computational exposition, which affords an immediate calculation of $J_1(h,f)$ at the singular set.

### 2.3.2 The correction

The global distribution $J_o(f)$ for a split regular $o$ has already been defined, by means of an orbital integral weighted by $v_0(g)$. It does not extend to the singular set of $A_0(\mathbb{A})$. As in the case of GL(2), to see how to introduce the corrected $J(h,f)$ it will be useful to calculate $J_o(f)$ explicitly. For that purpose we shall calculate $v_0(g)$. Write $g = bk$ with $b$ in $B(\mathbb{A})$ and $k$ in K. Since $H_0$ is right K-invariant we have $v_0(g) = v_0(b)$. To calculate $v_0(b)$ put

$$b = \begin{pmatrix} a_1 & n_1 & n_2 \\ & a_2 & n_3 \\ & & a_3 \end{pmatrix}, \qquad \alpha = (x,y,z) \text{ in } \mathbb{R}^3.$$

Note that the products in the denominators in the definition of $v_0(g)$ are taken over $\delta = \alpha_1$ and $\alpha_2$. Using the comment at the end of 2.1.3 and noting that the terms corresponding to $w_\alpha$, $\alpha$ equals id, $\alpha_1\alpha_2\alpha_1$, or $\alpha_1,\alpha_1\alpha_2$, or $\alpha_2,\alpha_2\alpha_1$, have (each pair) the same denominators, we obtain that $v_0(b) = \frac{3}{2}V_0$. Here $V_{2-i}$ is defined by the sum of

$$[(x\log\|a_1\|+y\log\|a_2\|+z\log\|a_3\|)^i + (z\log E+(y-z)\log A+(x-y)\log B)^i]/(x-y)(y-z),$$

$$[(y\log E+(x-y)\log C+(z-y)\log\|a_3\|)^i + (z\log E+(x-z)\log A+(y-x)\log D)^i]/(x-y)(z-x),$$

$$[((x-z)\log\|a_1\|+z\log E+(y-z)\log D)^i + (y\log E+(z-y)\log\|a_3\| C+(x-z)\log B)^i]/(y-z)(z-x).$$

We put

$$A = \|(a_1n_3, a_1a_2, n_1n_3-n_2a_2)\|, \qquad B = \|(a_1,n_1,n_2)\|,$$

$$C = \|(a_1,n_1)\|, \qquad D = \|(a_2,n_3)\|, \qquad E = \|a_1a_2a_3\|.$$

Since $V_1 = V_2 = 0$ the function $v_0(g) = v_0(b)$ is indeed left-invariant under $A_0(\mathbb{A})$. Hence we may assume that $a_1 = a_2 = a_3 = 1$, and find that $V_0 = \frac{2}{3}v_0(b)$ is equal to

$$2\log A\log B - (\log A/D)^2 - (\log B/C)^2. \tag{4}$$

In particular it is indeed independent of $(x,y,z)$.

Let $h=\operatorname{diag}(a,b,c)$ be a diagonal element of $\mathcal{O}$. The global function $J_{\mathcal{O}}(f)$ is given by:

$$J_{\mathcal{O}}(f) = \ell\int f^K(n^{-1}hn)v_0(n)dn = \ell\int f^K(hm)v_0(n)dm \qquad \left(n = \begin{pmatrix}1 & n_1 & n_2\\ & 1 & n_3\\ & & 1\end{pmatrix}, m = \begin{pmatrix}1 & m_1 & m_2\\ & 1 & m_3\\ & & 1\end{pmatrix}\right).$$

In the last integral $n$ is defined by $m = h^{-1}n^{-1}hn$, hence $n_1 = m_1/(1-b/a)$,

$$n_2 = (m_2-m_1m_3/(1-a/b))/(1-c/a), \qquad n_3 = m_3/(1-c/b).$$

As usual we put $f^K(g) = \int f(k^{-1}gk)dk\,(k \text{ in } K)$. It is clear that $v_0(n)$ has no finite limit as $h$ approaches the singular set of $A_0(\mathbb{A})$.

In an attempt to make $v_0(n)$ and $J_o(f)$ continuous, we shall multiply the $n_i$ throughout by their denominators. Thus we shall multiply the three entries of $A$ by $(1-c/a)(1-c/b)$, of $B$ by $(1-a/b)(1-c/a)$, of $C$ by $1-a/b$ and of $D$ by $1-c/b$. We can now introduce

$$J(h,f) = \frac{1}{4}\int f^K(hn)v_0(n)dn,$$

where $v_0(n)$ is defined by (4) and

$$A = \|((1-c/a)(1-c/b),\ n_3(1-c/a),\ n_1n_3 - n_2(1-c/b))\|, \qquad C = \|(1-a/b, \tfrac{a}{b}n_1)\|,$$

$$B = \|(1-c/a)(1-a/b),\ n_1(1-c/a)a/b,\ n_1n_3-n_2(1-a/b))\|, \qquad D = \|(1-c/b, n_3)\|.$$

Since $1-b/a, \ldots$ are F-rational and non-zero the product formula on $F$ implies that

$$\sum J(h,f) = 6\ J(h,f) = J_o(f) \qquad (h \text{ in } \mathcal{O}\cap A(F))$$

It is not yet clear that $J(h,f)$ extends continously to all of $A_0(\mathbb{A})$. We may express $A$, $B$, $C$, $D$ as products over $v$ of the local analogues $A_v, \ldots$ . We claim that it suffices to take these four products over a fixed finite set of places, containing all $v$ with $f_v \neq f_v^0$. This set depends only on $f_v$ for all $v$ but $v_0$, a fixed place; it is independent of $f_{v_0}$ and $h$. The proof is the same as in the case of $GL(2)$: There are only a finite number of $h$ with $f^K(hn) \neq 0$, and then for almost all $v$ the quantities $1-s/t$, for any eigenvalues $s,t$ of $h$, for any $h$ in our finite set, are equal to $1$ in the valuation $v$. When $f_v = f_v^0$ for such $v$ we have $A_v = B_v = \ldots = 1$, so that the product over $v$ can be taken over a fixed finite set. It follows that $J(h,f)$ extends to $A_0(\mathbb{A})$ as a continous function. The local distributions $J_o(f_v)$ and $J(h,f_v)$ are defined to be the product of $\Delta_v(h)$ and the local version of the integrals which define the global distributions, when $h$ is regular.

An alternative definition of the corrected weighted orbital integrals $J(h,f)$ is given by replacing $H_0(wg)$ by $H_0(wg) - \tilde{H}_0(w\rho)$ in the definition of $v_0(g)$. Here $\rho = \rho(h)$ is the element of $G(\mathbb{A})$ defined by $h^{-1}\rho^{-1}h\rho = u$, where

$$u = \begin{pmatrix} 1 & 1 & 0 \\ 0 & 1 & 1 \\ 0 & 0 & 1 \end{pmatrix}; \qquad \text{thus } \rho = \begin{pmatrix} 1 & x^{-1} & \frac{b}{a}x^{-1}y^{-1} \\ 0 & 1 & z^{-1} \\ 0 & 0 & 1 \end{pmatrix}$$

where $x = 1 - b/a$, $y = 1 - c/a$, $z = 1 - c/b$. The global function $\tilde{H}_0$ is by definition the sum of the local functions $\tilde{H}_{ov}$. $\tilde{H}_{ov}(w\rho)$ is the function equal to $H_{ov}(w\rho)$ when $|x|_v, |y|_v, |z|_v < 1$ and $v$ is non-archimedean. If $w = w_\alpha$ and $\alpha$ is the element $\alpha_1, \alpha_2, \alpha_2\alpha_1, \alpha_1\alpha_2, \alpha_1\alpha_2\alpha_1$ of $W$ then $\tilde{H}_{ov}(w\rho)$ is given by the value of $H_{0v}$ at the diagonal matrix $(x,x^{-1},1)$, $(1,z,z^{-1})$, $(x,y,x^{-1}y^{-1})$, $(yz,y^{-1}z^{-1})$, $(yz,z^{-1}x,x^{-1}y^{-1})$, respectively. For the last two cases note that $1/xz - b/axy = 1/yz$ . The product formula shows again that for $h$ in $A_0(F)$ the value of the global function $\tilde{H}_0(w\rho)$ is 0. From this exposition it is clear that $\{X_P = w^{-1}\tilde{H}_0(w\rho)\}(P = w^{-1}P_0w)$ is $A_0$-orthogonal, and $\{\exp\langle z,X_P\rangle\}(z \text{ in } \mathbb{E}^3)$ is a $(G,A_0)$ - family in the sense of [3], p.36, hence that [3], Lemma 8.2, as well as the formal Corollary 11.3 there, is valid not only for $J_o(f)$ but also for the corrected $J(h,f)$. The part of [3], Corollary 11.3 which we need will be recorded in the context of $J(h,f)$ in Lemma 6 below.

### 2.3.3 Singular classes

The significance of the limits of $J(h,f)$ at the singular set of $A_0(\mathbb{A})$ cannot be appreciated before the following two lemmas are recorded.

LEMMA 3. The sum of $J_o(f)$ over the $o$ which contain $h = \mathrm{diag}(h_1, h_2, h_3)$ with exactly two equal eigenvalues is equal to three times the product by $\ell$

<u>of the sum of</u>: (1) <u>the sum over</u> h <u>with</u> $h_1 = h_3$ <u>in</u> $A_0(F)$ <u>modulo</u> NZ(E) <u>of</u>

$$\tfrac{1}{4}\int f^K(hn)[2\sum\log A_v \sum\log C_vD_v - (\sum\log C_v)^2 - (\sum\log D_v)^2]dn$$

<u>and</u>

$$\tfrac{1}{2}(\lambda_0/\lambda_{-1} - \sum L_v'(1)/L_v(1)) \int f^K(hn) \sum\log C_vD_v dn \qquad (n \text{ in } N_0(\mathbb{A})).$$

<u>where</u>

$$A_v = |n_1n_3-n_2(1-h_1/h_2)|_v, \qquad C_v = \|(1-h_2/h_1,n_1)\|_v, \qquad D_v = \|(1-h_1/h_2,n_3)\|_v,$$

<u>and</u> (2) <u>the sum over</u> h <u>with</u> $h_1 = h_2$ <u>in</u> $A_0(F)$ <u>modulo</u> NZ(E) <u>of</u>

$$\tfrac{1}{2}|M_1(F)A_1(\mathbb{A})\backslash M_1(\mathbb{A})|\int_{K\cap M_1\backslash K}\int_{N_1(\mathbb{A})} f(k^{-1}hnk) \sum\log\|(1-h_3/h_1,n_2,n_3)\|_v dndk.$$

The sums over v extend over any fixed set of places which contain the v with $f_v \neq f_v^0$ and the v with $|1-h_2/h_1|_v \neq 1$ (for (1)) and $|1-h_3/h_1| \neq 1$ (for (2)). If all components $f_v$ of f, except perhaps $f_{v_0}$, are fixed, then there are only a finite number of h (in a fixed set independent of $f_{v_0}$) for which $f^K(hn) \neq 0$ for any n. The first line of (1) is clearly the value of J(h,f) at our h. The Lemma will be proved in section 2.4.

LEMMA 4. <u>The sum of</u> $J_o(f)$ <u>over the</u> $o$ <u>which contain scalars is equal to the product by</u> $\ell$ <u>of the sum over</u> h <u>in</u> Z(F) <u>modulo</u> NZ(E) <u>of the sum of</u> (1) <u>the integral over</u> n <u>in</u> $N_0(\mathbb{A})$ <u>of the product of</u> $f^K(hn)$ <u>and the sum of</u>

$$\sum\log|n_1|_v \sum\log|n_3|_v + \tfrac{1}{4}(\sum\log|n_1|_v)^2 + \tfrac{1}{4}(\sum\log|n_3|_v)^2,$$

$$\tfrac{3}{2}(\lambda_0/\lambda_{-1} - \sum L_v'(1)/L_v(1)) \sum \log|n_1n_3|_v,$$

$$\frac{3}{2}\sum_v \frac{L_v'(1)}{L_v(1)}\left(\sum_{w\neq v}\frac{L_w'(1)}{L_w(1)} - \frac{2\lambda_0}{\lambda_{-1}}\right) + \left(\frac{\lambda_0}{\lambda_{-1}}\right)^2 + \frac{\lambda_1}{\lambda_{-1}} + \sum L_v(1)[\tfrac{1}{2}(L_v(s)^{-1})''_{s=1} - (L_v(s)^{-1})'_{s=1}]$$

and (2)

$$3\int_{\mathbb{A}^\times} f^K\left(h\begin{pmatrix}1 & & c\\ & 1 & \\ & & 1\end{pmatrix}\right)|c|^2\log|c|\,d^\times c,$$

and (3)

$$|G(F)Z(\mathbb{A})\backslash G(\mathbb{A})|\,f(h).$$

All sums over $v$ in (1) are taken over any fixed finite set of places which includes all $v$ with $f_v \neq f_v^0$. It is clear that the limit of $J(h,f)$ at a scalar $h$ is the integral over $N_0(\mathbb{A})$ of the product of $f^K(hn)$ by the first line of (1). The Lemma will be proved in section 2.5.

To summarize the discussion, we state:

LEMMA 5. The sum of $J_o(f)$ over the split (regular or not) $o$ is equal to the sum of

$$\ell\sum J(h,f) \qquad (\text{all } h \text{ in } A_0(F)),$$

where $J(h,f)$ is the corrected weighted orbital integral of $f$ at $h$ defined in 2.3.2, the terms described by the second displayed line of (1) and by (2) in Lemma 3, and by the second and third displayed lines in (1), by (2) and by (3) in Lemma 4.

### 2.3.4 The term $\sum_o I_o(f)$

Lemma 5 expresses the sum of $J_o(f)$ over the $o$ which intersect $A_0(F)$ non-trivially in terms of the $A_0(F)$-values of the global regular function $J(h,f)$ of $h$ in $A_0(\mathbb{A})$. In sections 2.7 and 2.8 we shall show that the summation formula can be applied to $J(h,f)$ and the pair $A_0(F)$, $A_0(\mathbb{A})$. For the comparison of the trace formula with the twisted formula of the next chapter we have to (1) note that the camparison will have to be done in terms of the local components $f_v$ of $f$, in fact in terms of the (invariant) orbital integrals, (2) take into account the non-invariant part $X$ of $\sum J_\chi(f)$, the contribution to the trace formula from the continous spectrum.

The invariant terms in $\sum_o J_o(f)$ are those described in Lemma 1, (2) of Lemma 3, (3) of Lemma 4. The difference between the non-invariant part of $\sum_o J_o(f)$ and that $X$ of $\sum_\chi J_\chi(f)$ is clearly invariant, since all other terms in the trace formula are. To express it in terms of a sum over quadratic and split $o$ of products of local invariant distributions, Arthur [3] introduced invariant distributions $I_o(f)$ (globally) and $I_o(f_v)$ (locally), such that the sum over these $o$ of $I_o(f)$ is the required difference. $I(h,f) = I_o(f)$ is defined to be the difference between $J(h,f_v) = J_o(f_v)$ and some smooth function, compactly supported modulo $Z_E(\mathbb{A})$, when $h$ (and $o$) is quadratic; locally $I(h,f_v) = I_o(f_v)$ is defined by the difference of $J(h,f_v) = J_o(f_v)$ and a local function which, in addition to the above properties, vanishes if $f_v$ is spherical. The sum of $I_o(f)$ over the quadratic $o$ is described by Lemma 2 with $I(ah,f_v)$ replacing $J(ah,f_v)$, by virtue of [3], Cor. 11.3.

For split $o$ the invariant distribution $I_o(f)$ is defined in [3] by

$$I_o(f) = J_o(f) - J_{1o}(\Phi_1(f)) - \Phi_{0o}(f).$$

We also put

$$I(h,f) = J(h,f) - J_1(h,\Phi_1(f)) - \Phi_0(f,h).$$

Here $\Phi_1(f)$ is a smooth compactly supported function on $M_1(\mathbb{A})$ modulo $Z_E(\mathbb{A})$, which is the sum over $v$ where $f_v$ is not spherical (a finite set) of products over all places, where all factors except the one at $v$ is $f^K_{vN_1}$. If $h = \begin{pmatrix} h' & \\ & h'' \end{pmatrix}$ lies in $M_1(F_v)$ with $h''$ in $F_v^\times$, we put

$$f^K_{vN_1}(h) = |\det h'/h''^2|^{\frac{1}{2}} \int_{K\cap M_1\setminus K} \int_{N_1} f_v(k^{-1}hnk)\,dn\,dk.$$

If $o$ intersects $M_1$ in the classes $o_i$ of $M_1$ then $J_{1o}$ is the sum over $o_i$ of the distributions $J_{1o_i}$ on $M_1$; the latter distributions are defined in analogy with those of 2.3.1 on GL(2). $\Phi_0(f)$ is a smooth compactly supported function on $A_0(\mathbb{A})$ modulo $Z_E(\mathbb{A})$, and $\Phi_{0o}(f)$ is the sum of the values of $\Phi_0(f)$ at the points in the intersection of $o$ with $A_0$.

The definition of the local distributions $I_o(f_v)$ is analogous, with an additional condition: For spherical $f_v$ both $\Phi_1(f_v)$ and $\Phi_0(f_v)$ are 0. A further invariant distribution $I_{1o}(f_v)$ on $M_1(F_v)$ was defined in [3] as the difference of $J_{1o}(f_v)$ and the sum of the values of some smooth compactly supported (module $NZ(E_v)$) function on $A_0(f_v)$ at the points in the intersection of $o$ and $A_0(F)$. For the next lemma we shall need the corrected version $I_1(h,f_v)$ of $I_{1o}(f_v)$, in whose definition appears $J_1(h,f_v)$(not $J_{1o}(f_v)$), and the value of the subtracted function at h. Replacing $J(h,f_v)$ and $J_1(h,f_v)$ by the suitable average we may assume that they are constant on the intersection of $o$ and $A_0(F)$. Finally note that the index of $Z_E(\mathbb{A})$ in $Z(\mathbb{A})$ is $\ell$ and $dk = \lambda^2_{-1} \otimes dk_v$ for measures on K.

Although the following was stated in [3], Cor. 11.3 for the $I_o(f)$, as noted at the end of 2.3.2 its proof and statement are valid for the $I(h,f)$, which are the disbributions of use to us.

LEMMA 6. <u>For any regular split</u> h <u>in</u> $A_0(\mathbb{A})$ <u>the global distribution</u> $I(h,f)$ <u>is the sum of</u>

$$\ell\,\lambda^2_{-1} \sum_{v_1\neq v_2} \prod_{i=1,2} I_i(h,f^K_{v_iN_1}) \prod_{w\neq v_1,v_2} F(h,f_w)$$

<u>and</u>

$$\ell\,\lambda^2_{-1} \sum_{v} I(h,f_v) \prod_{w\neq v} F(h,f_w).$$

For reasons explained in 2.3.2 the sums over $v$, $v_1$, $v_2$ are to be taken over a fixed finite set, independent of $h$, containing all $w$ with $f_w \neq f_w^0$, which depends only on the support of the $f_w$ for $w \neq v_0$ for a fixed place $v_0$, or any larger set.

With the exception of the proofs of Lemmas 3 and 4 and the description of the asymptotic behaviour of $J(h,f_v)$, the discussion of the sum $\sum_o I_o(f)$ is finished.

LEMMA 7. The sum of $I_o(f)$ over all of the classes $o$ is equal to the sum of (1) the terms of Lemma 1, (2) the terms of Lemma 2 with $J(ah,f_v)$ replaced by $I(ah,f_v)$, (3) the sum over all $h$ in $A_0(F)$ modulo $NZ(E)$ of the two terms in Lemma 6, (4) the terms of the second displayed line of (1), and of (2), in Lemma 3, and those of the second and third displayed lines in (1), of (2), and of (3), in Lemma 4, and finally (5) the sum over all singular $h$ in $A_1(F)$ modulo $NZ(E)$ of $\delta(h)(= 1$ if $h$ lies in $Z(F)$, $= \frac{1}{3}$ otherwise) times

$$\ell(\lambda_0/\lambda_{-1} - \sum L_v'(1)/L_v(1))\Phi_1(f)^{K\cap M_1}_{N_0\cap M_1}(h) + \ell|M_1(F)A_1(\mathbb{A})\backslash M_1(\mathbb{A})|\Phi_1(f)(h).$$

The terms of (5) originate from $J_{1o}(f)$, with $\Phi_1(f)$ replacing $f$ and $o$ the class of a singular element in $M_1(F)$, which was described in 2.3.1.

Note that $\Phi_1(f)^{K\cap M_1}_{N_0\cap M_1}$ is the sum over $v$ where $f_v$ is not spherical (a finite set) of products over all places $w$, where all factors except the one at $v$ is $f^K_{wN_0}$.

## 2.4.1 Proof of Lemma 3

To prove Lemma 3 recall that the distribution $J_o(f)$ is given by the constant term in $t_i = \frac{3}{2}\langle T,\mu_i\rangle$ $(i = 1,2)$ of (3), where in our context $Z(\mathbb{A})$ there is $Z_E(\mathbb{A})$ here. We may replace $Z_E(\mathbb{A})$ by the centre $Z(\mathbb{A})$ of $G(\mathbb{A})$ if we multiply the expression by the index $\ell$ of $Z_E(\mathbb{A})$ in $Z(\mathbb{A})$. We may assume that $h$ is a multiple of diag(1,h,1) with $h \neq 1$ in $F^\times$, by a scalar in $F^\times/NE^\times$. For brevity we shall calculate $J(h,f)$ assuming that this scalar is equal to 1, although the calculation applies with any such scalar and is recorded in Lemma 3 in the general form.

To calculate (3) we rewrite the sums over $P$ and over $\gamma$ as sums over conjugacy classes in the intersection of $\mathcal{O}$ with the Levi subgroup $M$. A list of pairs consisting of: (1) a representative in each conjugacy class, (2) the corresponding parabolic subgroup, is given by:

$$\text{I: } \begin{pmatrix} 1 & & 1 \\ & h & \\ & & 1 \end{pmatrix}, G; \qquad \text{I}_1\text{: } \begin{pmatrix} 1 & 1 & \\ & 1 & \\ & & h \end{pmatrix}, P_1; \qquad \text{I}_2\text{: } \begin{pmatrix} h & & \\ & 1 & 1 \\ & & 1 \end{pmatrix}, P_2;$$

II: diag(1,h,1), $P_1$; $\text{II}_1$: diag(1,h,1), $P_2$; $\text{II}_2$: diag(h,1,1), $P_0$;

$\text{II}_3$: diag(1,1,h), $P_0$; $\text{II}_4$: diag(1,h,1), $P_0$;

III: diag(1,1,h), G; $\text{III}_1$: diag(1,1,h), $P_1$; $\text{III}_2$: diag(h,1,1), $P_2$.

To simplify the notations put $u(x) = \begin{pmatrix} 1 & & x \\ & h & \\ & & 1 \end{pmatrix}$.

The contribution from the classes I, $\text{I}_1$, $\text{I}_2$ is given by

$$\int f(g^{-1}u(1)g)[1-\hat{\tau}_1(H(w_{\alpha_2}g)-T) - \hat{\tau}_2(H(w_{\alpha_1}g)-T)]dg \tag{5}$$

with $g$ in $A_3(F)(N_0 \cap M_3)(F)Z(\mathbb{A})\backslash G(\mathbb{A})$, where

$$A_3 = \{\mathrm{diag}(a,b,a)\}, \; M_3 = \{\begin{pmatrix} * & 0 & * \\ 0 & * & 0 \\ * & 0 & * \end{pmatrix} \text{ in } G\}, \; N_3 = \{\begin{pmatrix} 1 & * & 0 \\ & 1 & * \\ & & 1 \end{pmatrix} \text{ in } N_0\}.$$

$N_3$ is not a subgroup of $N_0$, but a set of representatives for $N_0/N_1 \cap N_2$. We shall use the decomposition

$$g(=ank) = \begin{pmatrix} 1 & & n_2 \\ & 1 & \\ & & 1 \end{pmatrix}\begin{pmatrix} a & & \\ & b & \\ & & c \end{pmatrix}\begin{pmatrix} 1 & n_1 & 0 \\ & 1 & n_3 \\ & & 1 \end{pmatrix}k,$$

where $a, b, c$ in $\mathbb{A}^\times$, $n_1, n_2, n_3$ in $\mathbb{A}$, $k$ in $K$, and the modular function $|c/a|$. With the change $c \longrightarrow c/a$ the integrand with respect to $a$ becomes

$$\int_{N_3(\mathbb{A})} f^K(n_3^{-1}u(a)n_3)\int_{F^\times\backslash\mathbb{A}^\times} A(a,b,c;n_1,n_3)\ d^\times(b/c)|a|dn_3,$$

where

$$A(a;b,c;n_1,n_3) = 1 - \hat{\tau}_1[(\log|c/a|,\ \log|c|,\ \log|b|) + (0,\ -\log\|(1,n_3)\|,\ \log\|(1,n_3)\|) - T]$$

$$- \hat{\tau}_2[(\log|b|,\ \log|c/a|,\ \log|c|) + (-\log\|(1,n_1)\|,\ \log\|(1,n_1)\|,\ 0) - T]$$

Since $t_i = \frac{3}{2}\langle T,\mu_i\rangle > 0$ $(i = 1,2)$, $A(a,b,c;n_1,n_3)$ is $0$ unless

$$- t_2 - \tfrac{3}{2}\log\|(1,n_3)\| - \tfrac{1}{2}\log|a| < \log|b/c| < t_1 + \tfrac{3}{2}\log\|(1,n_1)\| - \tfrac{1}{2}\log|a|,$$

where it is 1. Integrating with respect to $b/c$ we obtain $F(a)|a|$, where

$$F(a) = \int_{N_3(\mathbb{A})} f^K(n_3^{-1}u(a)n_3)\ (t_1+t_2+\tfrac{3}{2}\log\|(1,n_1)\| + \tfrac{3}{2}\log\|(1,n_3)\|)dn_3.$$

The integral over $a$ will be taken later.

2.4.2 The contribution from the Classes $II, \ldots, II_4$ is

$$- \iint_{\mathbb{A}} f^K(g^{-1}u(n)g)[\hat{\tau}_1(H(w_{\alpha_1}g)-T) + \hat{\tau}_2(H(w_{\alpha_2}g)-T)$$

$$- \hat{\tau}_0(H(w_{\alpha_1}g)-T) - \hat{\tau}_0(H(w_{\alpha_2}g)-T) - \hat{\tau}_0(H(g)-T)]dndg,$$

where $g$ is taken over $A_0(F)(N_0\cap M_3)(F)Z(\mathbb{A})\backslash G(\mathbb{A})$. Using the above decomposition and change of variables the integral with respect to $a$ becomes

$$- \int_{N_3(\mathbb{A})}\int_{\mathbb{A}} f^K(n_3^{-1}u(n)n_3)\int_{F^\times\backslash\mathbb{A}^\times}(A_1+A_2-A_3)\ d^\times(b/c)dndn_3,$$

where

$$A_1 = (\hat{\tau}_1-\hat{\tau}_0)[(\log|b|, \log|c/a|, \log|c|) + (-\log\|(1,n_1)\|, \log\|(1,n_1)\|, 0)-T],$$

$$A_2 = (\hat{\tau}_2-\hat{\tau}_0)[(\log|c/a|, \log|c|, \log|b|) + (0, -\log\|(1,n_3)\|, \log\|(1,n_3)\|)-T],$$

$$A_3 = \hat{\tau}_0[(\log|c/a|, \log|b|, \log|c|)-T].$$

These $A_i$ obtain the values 0 or 1, and $A_1 \neq 0$ if

$$I_2 = \log|a| + 2t_2 < \log|b/c| < I_4 = -\tfrac{1}{2}\log|a| + \tfrac{3}{2}\log\|(1,n_1)\| + t_1,$$

$A_2 \neq 0$ if

$$I_1 = -\tfrac{1}{2}\log|a| - \tfrac{3}{2}\log\|(1,n_3)\| - t_2 < \log|b/c| < I_3 = -2\log|a| - 2t_1,$$

and $A_3 \neq 0$ if $I_2 < \log|b/c| < I_3$. Since (i) $I_1 < I_4$, (ii) if $I_2 < I_3$ then $I_1 < I_3$, $I_2 < I_4$, (iii) if $I_2 > I_3$ then $I_3 < I_4$, $I_1 < I_2$, then $A_i$ $(i = 1,2,3)$ are either all 0 or all non-zero, and the integral of $A_1 + A_2 - A_3$ with respect to $b/c$ is $\frac{3}{2}I$, where

$$I = \tfrac{2}{3}\max(0,I_3-I_2) - \tfrac{2}{3}\max(0,I_3-I_4) - \tfrac{2}{3}\max(0,I_1-I_2).$$

$$= -\min(0,2\log|a|+\tfrac{4}{3}(t_1+t_2)) + \min(0, \log|a|+\log\|(1,n_1)\|+2t_1) + \min(0, \log|a|+\log\|(1,n_3)\|+ 2t_2)$$

Clearly $I \neq 0$ only if $|a| < 1$.

To perform the integration with respect to $a$ we write each element in $\mathbb{A}^\times$ as the product $xa$ of $x$ in $F^\times$ and $a$ in $F^\times\backslash\mathbb{A}^\times$, and apply the Poisson summation formula to the function $F(xa)$ of $a$ over the domain $|a| \leq 1$ in $F^\times\backslash\mathbb{A}^\times$. Denoting the Fourier transform of $F$ by $\hat{F}$ we obtain the sum of

$$\int_{|a|>1} |a|\sum_x F(xa)d^\times a + \int_{|a|\leq 1} \sum_x \hat{F}(a^{-1}x)d^\times a - \int_{|a|\leq 1} F(0)|a|d^\times a$$

and

$$\frac{3}{2}\int_{N_3(\mathbb{A})}\int_{\mathbb{A}} f^K(n_3^{-1}u(n)n_3)dn\int_{|a|\leq 1}(\tfrac{2}{3}(t_1+t_2)+\log\|(1,n_1)\|\,\|(1,n_3)\|-I)d^\times a\,dn_3$$

To integrate over $a$ note that the integrand is 0 if $\log|a|$ is less than $L_1 = -2t_2 - \log\|(1,n_3)\|$, $L_2 = -2t_1 - \log\|(1,n_1)\|$ and hence $L_3 = -\frac{2}{3}(t_1+t_2)$. If, for example, $L_1 \leq L_2 \leq L_3$, then the constant term of the integral over $a$ (which is a quadratic polynomial in $t_1, t_2$) is

$$\int_{L_1}^{L_2} -(\log\|(1,n_3)\|+x)dx + \int_{L_2}^{L_3} -(\log\|(1,n_1)\|\cdot\|(1,n_3)\|+2x)dx$$

$$= -\tfrac{1}{2}(\log\|(1,n_1)\|)^2 - \tfrac{1}{2}(\log\|(1,n_3)\|)^2.$$

2.4.3 If the power series expansion of a meromorphic function $F(s)$ at $s = 0$ takes the form $\ldots + a_{-1}s^{-1} + a_0 + a_1 s + \ldots$ then we write $\mathrm{f.p}_{s=0}F(s)$ for $a_0$. We proved that the contribution from the I's and the II's is equal to

$$\frac{3}{2}\int_{N_3(\mathbb{A})}\mathrm{f.p.}_{s=0}\Big\{\int_{\mathbb{A}^\times} f^K(n_3^{-1}u(a)n_3)\;|a|^{1+s}d^\times a\Big\}\log\|(1,n_1)\|\cdot\|(1,n_3)\|\,dn_3$$

$$-\frac{3}{4}\int_{N_3(\mathbb{A})}\int_{\mathbb{A}} f^K(n_3^{-1}u(n)n_3)\;[(\log\|(1,n_1)\|)^2 + (\log\|(1,n_3)\|)^2]dn\,dn_3.$$

In calculating the f.p. here again we express each element in $\mathbb{A}^\times$ as a product $xa$ with $x$ in $F^\times$ and $a$ in $F^\times\backslash\mathbb{A}^\times$, and apply the Poisson summation formula on the domain $|a| \leq 1$.

There is a better way to express this expression. We introduce the function

$$\theta(s,f) = \Pi_v \frac{|(1-h)(1-h^{-1})|_v}{L(1+s,1_v)} \int_{F_v^\times} f_v^{K_v}(n_3^{-1}u(a)n_3)|a|^{1+s}d^\times a.$$

For each $v$ with $|h|_v = 1$ and $f_v = f_v^0$ the local factor is just the product of the volume of $N_{3v} \cap K_v$ and that of the $a$ in $F_v^\times$ with $|a|_v = 1$ with respect to $d^\times a_v (=L(1,1_v)da_v/|a_v|)$. Since the product of these constants over almost all $v$ is convergent the function $\theta(s,f)$ is analytic in a neighborhood of $s = 0$ and can be differentiated term by term there. Our f.p. function is just the product of $\lambda_{-1}\theta(s,f)$ and

$$L(1+s,1_F) = \lambda_{-1}s^{-1} + \lambda_0 + \lambda_1 s + \ldots,$$

since $d^\times a = (\lambda_{-1})^{-1} \otimes d^\times a_v$ and $dk = (\lambda_{-1})^2 \otimes dk_v$. Hence the value of f.p. at $s = 0$ is $\lambda_{-1}^2\theta'(0,f) + \lambda_{-1}\lambda_0\theta(0,f)$.

An explicit global expression for the contribution form the I's and II's is obtained after dividing by $(\lambda_{-1})^2$:

$$\frac{3}{2}\iint f^K(n_3^{-1}u(n)n_3)$$

$$[\sum \log\|(1,n_1)\|_v\|(1,n_3)\|_v \sum \log|n|_v - \tfrac{1}{2}(\sum \log\|(1,n_1)\|_v)^2 - \tfrac{1}{2}(\sum \log\|(1,n_3)\|_v)^2]dndn_3$$

$$+ \frac{3}{2}(\lambda_0/\lambda_{-1} - L_v'(1)/L_v(1)) \iint f^K(n_3^{-1}u(n)n_3) \sum \log\|(1,n_1)\|_v\|(1,n_3)\|_v \, dndn_3$$

$$(n \text{ in } \mathbb{A},\ n_3 \text{ in } N_3(\mathbb{A})).$$

All sums are taken over any fixed finite set of places which contains all $v$ for which $f_v \neq f_v^0$. Changing $n_3^{-1}u(n)n_3$ to $u(0)n$ ($n$ in $N_0(\mathbb{A})$) this becomes

$$\frac{3}{2}\int f^K(u(0)n)[\sum \log A_v \sum \log C_vD_v - \tfrac{1}{2}(\sum \log C_v)^2 - \tfrac{1}{2}(\sum \log D_v)^2]dn$$

$$+\frac{3}{2}(\lambda_0/\lambda_{-1}-\sum L_v'(1)/L_v(1))\int f^K(u(0)n)\sum \log C_v D_v \, dn \qquad (n \text{ in } N_0(\mathbb{A}),$$

where

$$A_v = |n_1n_3-n_2(1-h^{-1})|_v, \qquad C_v = \|(1-h,n_1)\|_v, \qquad D_v = \|(1-h^{-1},n_3)\|_v.$$

The sums extend over any fixed finite set of places which contains all $v$ for which $f_v \neq f_v^0$ and all $v$ for which $|1-h|_v \neq 1$. Note that if all components $f_v$, except perhaps $f_{v_0}$, are fixed, then there are only a finite number of $h$(in a fixed set independent of $f_{v_0}$) for which $f^K(u(0)n) \neq 0$ for any $n$.

Noting that there are three elements in the intersection of $\mathcal{O}$ and $A_0(F)$ part (1) of Lemma 3 follows at once.

2.4.4 The final contribution comes from the classes III, $III_1$, $III_2$. It is

$$\int_{M_1(F)Z(\mathbb{A})\backslash G(\mathbb{A})} f(g^{-1}hg)[1-\hat{\tau}_1(H(g)-T)-\hat{\tau}_2(H(w_{\alpha_1\alpha_2}g)-T)]dg,$$

where for brevity we wrote $h$ for $\mathrm{diag}(1,1,h)$. Since $g = amnk$ with $a$ in $Z(\mathbb{A})A_1(F)\backslash A_1(\mathbb{A})$ and $m,n,k$ in $M_1,N_1,K\cap M_1\backslash K$, this becomes

$$\int_{K\cap M_1\backslash K}\int_{N_1(\mathbb{A})} f(k^{-1}n^{-1}hnk)\int_{A_1(F)Z(\mathbb{A})\backslash A_1(\mathbb{A})}\{1-\hat{\tau}_1(H_0(a)-T)$$

$$-\hat{\tau}_2[H_0(w_{\alpha_1\alpha_2}a)+(-\log\|(1,n_2,n_3)\|, \log\|(1,n_2,n_3)\|/\|(1,n_3)\|, \log\|(1,n_3)\|)-T]da\}dndk$$

multiplied by

$$\int_{M_1(F)A_1(\mathbb{A})\backslash M_1(\mathbb{A})} dm = |M_1(F)A_1(\mathbb{A})\backslash M_1(\mathbb{A})|.$$

Using the measure preserving transformation $a \longrightarrow H_1(a) = x\mu_2(x \text{ in } \mathbb{R})$ we see that the inner integral is non-zero only when

$$-\frac{3}{2}\log\|(1,n_2,n_3)\| - t_1 < x < t_2;$$

hence we get

$$|M_1(F)A_1(\mathbb{A})\backslash M_1(\mathbb{A})| \int_{K\cap M_1\backslash K}\int_{N_1(\mathbb{A})} f(k^{-1}n^{-1}hnk)\sum \log\|(1,n_2,n_3)\|.$$

The sum is over any finite set of places containing all $v$ with $f_v \neq f_v^0$. Making the change $h^{-1}n^{-1}hn \longrightarrow n$, the $n_i$ become $n_i/(1-h)$, and correcting as usual we may replace the weight factor by $\sum \log\|(1-h,n_2,n_3)\|$ where the sum contains in addition all $v$ with $|1-h|_v \neq 1$. This is (2) of Lemma 3.

2.5.1 Proof of Lemma 4

It remains to consider the classes $\mathcal{O}$ in (2) which contain an element $h$ of $Z(F)$ modulo $NZ(E)$. For notational reasons we assume that $h = 1$ in the following calculation; it is valid for any $h$, and is recorded in the general form in the Lemma. Again we replace $Z_E(\mathbb{A})$ (denoted by $Z(\mathbb{A})$ in (3)) by $Z(\mathbb{A})$ and multiply that expression by $\ell$. To evaluate the expression at $h = 1$ consider first the contribution form the conjugacy classes of

$$\begin{pmatrix} 1 & 1 & \\ & 1 & 1 \\ & & 1 \end{pmatrix} \text{ in } G, \quad \begin{pmatrix} 1 & 1 & \\ & 1 & \\ & & 1 \end{pmatrix} \text{ in } P_1, \quad \begin{pmatrix} 1 & & \\ & 1 & 1 \\ & & 1 \end{pmatrix} \text{ in } P_2, \quad \begin{pmatrix} 1 & & \\ & 1 & \\ & & 1 \end{pmatrix} \text{ in } P_0.$$

Denote by $\alpha, \gamma$ any elements of $F^\times$, by $\beta$ any element of $F$; the $m_i$ are taken in $\mathbb{A}$. We have

$$\int_{P_0(F)Z(\mathbb{A})\backslash G(\mathbb{A})} [\sum f\left(g^{-1}\begin{pmatrix}1&\alpha&\beta\\&1&\gamma\\&&1\end{pmatrix}g\right) - \sum\int_{\mathbb{A}^2} f\left(g^{-1}\begin{pmatrix}1&\alpha&m_2\\&1&m_3\\&&1\end{pmatrix}g\right)\hat{\tau}_1(H_0(g)-T)$$

$$-\sum\int_{\mathbb{A}^2} f\left(g^{-1}\begin{pmatrix}1&m_1&m_2\\&1&\gamma\\&&1\end{pmatrix}g\right)\hat{\tau}_2(H_0(g)-T) + \int_{N_0(\mathbb{A})} f(g^{-1}ng)\hat{\tau}_0(H_0(g)-T)dg.$$

We shall use the decomposition $g = n(a^{-1},1,c)k$ for $Z\backslash G$ (with $n$ in $N_0$, $k$ in $K$) and the modular function $|ac|^2$. We obtain

$$\int |ac|^2 \int [\sum f^K\begin{pmatrix}1&a\alpha&ac(\beta+\alpha n_3-\gamma n_1)\\&1&c\gamma\\&&1\end{pmatrix}$$

$$-\ \hat{\tau}_1((-\log|a|,0,\log|c|)-T)\sum\int f^K\begin{pmatrix}1&a\alpha&ac(m_2+n_3\alpha-n_1m_3)\\&1&cm_3\\&&1\end{pmatrix}$$

$$-\ \hat{\tau}_2((-\log|a|,0,\log|c|)-T)\sum\int f^K\begin{pmatrix}1&am_1&ac(m_2+n_3m_1-n_1\gamma)\\&1&c\gamma\\&&1\end{pmatrix}$$

$$+\ \hat{\tau}_0((-\log|a|,0,\log|c|)-T)\int f^K\begin{pmatrix}1&am_1&ac(m_2+n_3m_1-n_1m_3)\\&1&cm_3\\&&1\end{pmatrix}],$$

where the first integral is over $a,c$ in $F^\times\backslash\mathbb{A}^\times$, the second is over $n_1,n_2,n_3$ in $\mathbb{A}$ mod $F$, the sums are over $(\alpha,\beta,\gamma)$, $(\alpha,\beta)$, $(\beta,\gamma)$ and the inner integrals are over $(m_2,m_3)$, $(m_1,m_2)$ and $(m_1,m_2,m_3)$.

Put

$$F(x,y) = \int_{\mathbb{A}} f^K\begin{pmatrix}1&x&n\\&1&y\\&&1\end{pmatrix}dn,$$

and rewrite the above expression in the form

$$\int_{a,c} [\sum_{\alpha,\gamma} F(a\alpha,c\gamma)|ac| - \chi(|ac^2|<u_2^{-2})\sum_{\alpha} F(a\alpha,\hat{0})|a|$$

$$-\chi(|a^2c|<u_1^{-2})\ \sum_\gamma F(\hat{0},c)|c| + \hat{F}(0,0)\chi(|a^2c|<u_2^{-2})\chi(|ac^2|<u_1^{-2})]d^\times a d^\times c.$$

Here $t_i = \log u_i$ $(i = 1,2)$. We denote by $\chi(\ldots)$ the characteristic function of the domain defined by the conditions in brackets. $F(\hat{x},y)$, $F(x,\hat{y})$, and $\hat{F}(x,y)$ denote respectively the Fourier transform of $F$ with respect to the first, second or both variables.

2.5.2 We shall now decompose the domain of integration to 4 subdomains, and apply the Poisson summation formula with respect to the suitable variable, as follows. We have the integral over $a$ and $c$ of the sum of

$$\sum_{\alpha,\gamma} F(a\alpha,c\gamma)|ac|\chi(|a^2c|,|ac^2|>1),$$

$$[\sum_{\alpha,\gamma} F(a\alpha,c^{-\hat{1}}\gamma)|a| + (\sum_\alpha F(a\alpha,\hat{0})|a|)_* - \sum_\alpha F(a\alpha,0)|ac|]\chi(|ac^2|\le 1,\ |a^2c|>1),$$

$$[\sum_{\alpha,\gamma} F(a^{-\hat{1}}\alpha,c\gamma)|c| + (\sum_\gamma F(\hat{0},c\gamma)|c|)_{**} - F(0,c\gamma)|ac|]\chi(|a^2c|\le 1,\ |ac^2|>1),$$

$$[\sum_{\alpha,\gamma}\hat{F}(a^{-1}\alpha,c^{-1}\gamma) + (\sum_\alpha \hat{F}(a^{-1}\alpha,0))_* + (\sum_\gamma \hat{F}(0,c^{-1}\gamma))_{**} - \sum_\alpha F(a^{-\hat{1}}\alpha,0)|c| - \sum_\gamma F(0,c^{-\hat{1}}\alpha)|a|$$

$$- (F(0,\hat{0})|a|)_* - (F(\hat{0},0)|c|)_{**} + F(0,0)|ac| + \hat{F}(0,0)]\chi(|a^2c|,|ac^2|\le 1),$$

(up to here we have rewritten only $\sum F(a\alpha,c\gamma)|ac|$), and

$$-\sum_\alpha (F(a\alpha,\hat{0})|a|)_*\chi(|ac^2|\le u_2^{-2},\ |ac^2|>1)$$

$$-[(\sum_\alpha \hat{F}(a^{-1}\alpha,0))_* - (F(0,\hat{0})|a|)_* + \hat{F}(0,0)]\chi(|ac^2|\le u_2^{-2},\ |a^2c|\le 1),$$

$$-\sum_\gamma (F(\hat{0},c\gamma)|c|)_{**}\chi(|a^2c| \leq u_1^{-2},\ |ac^2| > 1)$$

$$-\ [(\sum_\gamma \hat{F}(0,c^{-1}\gamma))_{**} - (F(\hat{0},0)|c|)_{**} + \hat{F}(0,0)]\chi(|a^2c| \leq u_1^{-2},\ |ac^2| \leq 1),$$

and finally

$$\hat{F}(0,0)\chi(|a^2c| \leq u_1^{-2}, |ac^2| \leq u_2^{-2}).$$

The integrals involving $\hat{F}(0,0)$ add up to $\hat{F}(0,0)$ times the homogeneous polynomial of degree 2 in $t_1, t_2$ given by $\iint dxdy$(over $-2t_2 < x + 2y < 0$, $-2t_1 < 2x + y < 0$), which can be ignored since we are interested only in the constant term in $t_1$ and $t_2$.

Taking together all of the terms indexed by * we get (put $u = u_2$)

$$\int_{u^{-2} < |ac^2| \leq 1} [\int_{|a^2c| > 1} \sum_\alpha F(a\alpha,\hat{0})|a|$$

$$+ \int_{|a^2c| \leq 1} (\sum_\alpha \hat{F}(a^{-1}\alpha,0) - |a|F(0,\hat{0}))]d^\times a d^\times c.$$

Changing $c$ to $ca^{-2}$ this becomes

$$\int_{|a|^{3/2}u^{-1} \leq |c| \leq |a|^{3/2}} (\int_{|c| > 1} \cdots \int_{|c| \leq 1} \cdots),$$

which is $$\log u \int_{|a| \leq 1} (\sum_\alpha \hat{F}(a^{-1}\alpha,0) - |a|F(0,\hat{0}))d^\times a$$

$$+ \int_{1 < |a| \leq u^{2/3}} (\log t - \tfrac{3}{2}\log|a|)\ (\sum_\alpha \hat{F}(a^{-1}\alpha,0) - |a|F(0,\hat{0}))d^\times a$$

$$+ \int_{1 < |a| \leq u^{2/3}} \tfrac{3}{2} \log|a| \sum_\alpha F(a\alpha,\hat{0})|a| d^\times a$$

$$+ \log u \int_{u^{2/3} \leq |a|} \sum_\alpha F(a\alpha,\hat{0})|a| d^\times a.$$

Applying the Poisson summation formula to the second summand here we get

$$\int_{1 < |a| \leq u^{2/3}} (\log u - \tfrac{3}{2}\log|a|)[\textstyle\sum_\alpha F(a\alpha,\hat{0})|a| - \hat{F}(0,0)]d^\times a$$

$$= \log u \int \textstyle\sum_\alpha F(a\alpha,\hat{0})|a|d^\times a - \log u \cdot \hat{F}(0,0)\cdot\tfrac{2}{3}\log u$$

$$- \tfrac{3}{2}\int \log|a| \textstyle\sum_\alpha F(a\alpha,\hat{0})|a|d^\times a + \tfrac{3}{2}\int \log|a|\hat{F}(0,0)d^\times a.$$

The third term here is just the negative of the third term in the previous expression. The first term here together with the fourth above is

$$\log u \int_{1 < |a|} \textstyle\sum_\alpha F(a\alpha,\hat{0})|a|d^\times a.$$

The fourth term here is obviously a scalar multiple of $(\log u)^2$ or $t_2{}^2$.

Similar punishment can be given to the terms indexed by **. The conclusion is that the constantterm with respect to $t_1$ and $t_2$ involves only the non-indexed (by * or **) and free of $\hat{F}(0,0)$ terms in (the first four lines of) our expression.

2.5.3 A more useful expression for the above constant term will be obtained on considering the following meromorphic function in the complex variables s, s':

$$\int_{\mathbb{A}^\times}\int_{\mathbb{A}^\times} F(a,c)|a|^s\ |c|^{s'}\ d^\times a d^\times c = \int_{F^\times\backslash\mathbb{A}^\times}\int_{F^\times\backslash\mathbb{A}^\times} \textstyle\sum_{\alpha,\gamma} F(a\alpha,c\gamma)|a|^s\ |c|^{s'}\ d^\times a d^\times c.$$

Decomposing the domain and applying the Poisson formula we get the integral of the sum of

(i) $$\textstyle\sum_{\alpha,\gamma} F(a\alpha,c\gamma)|a|^s\ |c|^{s'}\ \chi(|ac^2|,|a^2c| > 1),$$

(ii) $$[\sum_{\alpha,\gamma} F(a\alpha, c^{-1}\hat{\gamma})|a|^{s}|c|^{s'-1} + (\sum_\alpha F(a\alpha,\hat{0})|a|^{s}|c|^{s'-1})_*$$
$$- \sum_\alpha F(a\alpha,0)|a|^{s}|c|^{s'}]\chi(|ac^2| \leq 1,\ |a^2c| > 1),$$

(iii) $$[\sum_{\alpha,\gamma} F(a^{-1}\hat{\alpha}, c\gamma)|a|^{s-1}|c|^{s'} + (\sum_\gamma F(\hat{0},c\gamma)|a|^{s-1}|c|^{s'})_{**}$$
$$- \sum_\gamma F(0,c\alpha)|a|^{s}|c|^{s'}]\chi(|a^2c| \leq 1,\ |ac^2| > 1),$$

and finally (iv) the product of $\chi(|a^2c|,|ac^2| \leq 1)$ and

$$\sum_{\alpha,\gamma} \hat{F}(a^{-1}\alpha, c^{-1}\gamma)|a|^{s-1}|c|^{s'-1} + (\sum_\alpha \hat{F}(a^{-1}\alpha,0)|a|^{s-1}|c|^{s'-1})_*$$
$$+ (\sum_\gamma \hat{F}(0,c^{-1}\gamma)|a|^{s-1}|c|^{s'-1})_{**}$$
$$- \sum_\alpha F(a^{-1}\hat{\alpha},0)|a|^{s-1}|c|^{s'} - \sum_\gamma F(0,c^{-1}\hat{\gamma})|a|^{s}|c|^{s'-1}$$
$$- (F(0,\hat{0})|a|^{s}|c|^{s'-1})_* - (F(\hat{0},0)|a|^{s-1}|c|^{s'})_{**}$$
$$+ F(0,0)|a|^{s}|c|^{s'} + (\hat{F}(0,0)|a|^{s-1}|c|^{s'-1})_{***}.$$

The non-indexed terms are regular on Re $s_i \geq 1$ $(i = 1,2)$, and equal to our constant term. Those indexed by * give

$$\int_{|ac^2| \leq 1} [\int_{|a^2c| > 1} \sum_\alpha F(a\alpha,\hat{0})|a|^{s}|c|^{s'-1}$$
$$+ \int_{|a^2c| \leq 1} (\sum_\alpha \hat{F}(a^{-1}\alpha,0)|a|^{s-1}|c|^{s'-1} - F(0,\hat{0})|a|^{s}|c|^{s'-1})]d^\times a d^\times c$$

when Re $s_i > 1$. Applying Poisson's formula to the variable $a$ and the second integral we get

$$\int d^\times a[(\int_{|c| \leq |a|^{-1/2}} |c|^{s'-1} d^\times c) \sum_\alpha F(a\alpha,\hat{0})|a|^{s}$$

$$- \int_{|c| \leq |a|^{-2},\ |a|^{-1/2}} \hat{F}(0,0)|a|^{s-1}|c|^{s'-1} d^{\times}c].$$

This is equal to $(s'-1)^{-1}$ times the difference between

$$\int \sum_{\alpha} F(a\alpha, \hat{0})|a|^{s-1/2s'+1/2} d^{\times}a$$

and $\hat{F}(0,0)$ times

$$\int_{|a|>1} |a|^{s-1-2(s'-1)} d^{\times}a + \int_{|a|\leq 1} |a|^{s-1-1/2(s'-1)} d^{\times}a$$

$$= [2(s'-1) - (s-1)]^{-1} + [s-1-\tfrac{1}{2}(s'-1)]^{-1}.$$

The last equality holds when $2\,\mathrm{Re}(s'-1) > \mathrm{Re}(s-1) > \frac{1}{2}\mathrm{Re}(s'-1)$. Similar considerations apply to the terms indexed by $_{**}$. The term indexed by $_{***}$ is equal to

$$-3\,\hat{F}(0,0)/(2s'-s)(2s-s').$$

2.5.4 We showed that the constant term under consideration is equal to the value at the origin of the following function (which is analytic at $(s,s') = (0,0)$):

$$L(1+s)L(1+s')\theta(s,s') - \lambda_{-1}s^{-1}L(1+s-\tfrac{1}{2}s')\theta_1(s-\tfrac{1}{2}s') - \lambda_{-1}s'^{-1}L(1+s'-\tfrac{1}{2}s)\theta_2(s'-\tfrac{1}{2}s)$$
$$+ 3\lambda_{-1}^{2}\theta(0,0)/(2s-s')(2s'-s).$$

Here $L(s) = L(s,1_F)$ and below $L_v(s) = L(s,1_v)$, $n = \begin{pmatrix} 1 & a & b \\ & 1 & c \\ & & 1 \end{pmatrix}$, $b$ in $F_v$; then

$$\theta(s,s') = \prod_v (L_v(1+s)L_v(1+s'))^{-1} \iiint f_v^{K}(n)|a|^{1+s}|c|^{1+s'} d^{\times}a\,db\,d^{\times}c \quad (a,c \text{ in } F_v^{\times}),$$

$$\theta_1(s) = \prod_v L_v(1+s)^{-1} \iiint f_v^{K}(n)|a|^{1+s} d^{\times}a\,db\,dc \quad (a \text{ in } F_v^{\times},\ c \text{ in } F_v),$$

$$\theta_2(s) = \prod_v L_v(1+s)^{-1} \iiint f_v^{K}(n)|c|^{1+s} da\,db\,d^{\times}c \quad (a \text{ in } F_v,\ c \text{ in } F_v^{\times}).$$

Note that the relations $d^{\times}a = (\lambda_{-1})^{-1}\otimes d^{\times}a_v$ and $dk = (\lambda_{-1})^{2}\otimes dk_v$ between local and global measures on $\mathbb{A}^{\times}$ and $F_v^{\times}$, and K, were used; for example $\hat{F}(0,0) = (\lambda_{-1})^{2}\theta(0,0)$. Further, note that if $f_v = f_v^0$ then the corresponding factor of $\theta, \theta_1$ and $\theta_2$ is just the third power of the volume of $|x| \leq 1$ by $dx$ (=volume of $|x| = 1$ by $d^{\times}x$), as long as $\operatorname{Re} s_i > -1$. The product of these numbers over almost all $v$ is convergent; hence the products which define $\theta, \theta_1$ and $\theta_2$ are absolutely convergent and can be differentiated term by term.

Recall that our function, which is explicitly given by

$$\sum_{i=-1}^{\infty} \lambda_i s^i \sum_{j=-1}^{\infty} \lambda_j s'^j \sum_{k,\ell=0}^{\infty} k!^{-1}\ell!^{-1}\partial_1^k\partial_2^\ell \theta\, s^k s'^\ell \;+\; 3\lambda_{-1}^2\theta/(2s'-s)(2s-s')$$

$$-\lambda_{-1}s'^{-1}\sum_{i=-1}^{\infty}\lambda_i(s-\tfrac{1}{2}s')^i\sum_{k=0}^{\infty}k!^{-1}\partial_1^k\theta(s-\tfrac{1}{2}s')^k \;-\; \lambda_{-1}s^{-1}\sum_{j=-1}^{\infty}\lambda_j(s'-\tfrac{1}{2}s)^j\sum_{\ell=0}^{\infty}\ell!^{-1}\partial_2^\ell\theta(s'-\tfrac{1}{2}s)^\ell$$

where $\theta = \partial_1^0\partial_2^0\theta$ and $\partial_1^k\partial_2^\ell\theta$ denotes the value at $(s,s') = (0,0)$ of the $k$-th derivative of $\theta(s,s')$ with respect to $s$ and the $\ell$-th with respect to $s'$, is analytic at the origin. Its value there is the same as the value at $0$ of the analytic (at 0) function (of $s$) obtained by setting $s' = s$. It is

$$\sum_{i+j+k+\ell=0}\lambda_i\lambda_j\partial_1^k\partial_2^\ell\theta/k\ell \;-\; \tfrac{1}{2}\lambda_{-1}\Big[\sum_{i+k=1}\lambda_i\partial_1^k\theta/k + \sum_{j+\ell=1}\lambda_j\partial_2^\ell\theta/\ell\Big]$$

$$= \lambda_{-1}^2[\partial_1\partial_2\theta+\tfrac{1}{4}(\partial_1^2\theta+\partial_2^2\theta)] + \frac{3}{2}\lambda_0\lambda_{-1}(\partial_1\theta+\partial_2\theta) + (\lambda_0^2+\lambda_1\lambda_{-1})\theta.$$

Noting the disappearance of $\lambda_{-1}^2$ an explicit global form of this expression is given by the integral over $n$ in $N_0(\mathbb{A})$ of the product of $f^K(n)$ and the sum of

$$\sum \log|n_1|_v \sum \log|n_3|_v + \tfrac{1}{4}(\sum \log|n_1|_v)^2 + \tfrac{1}{4}(\sum \log|n_3|_v)^2,$$

$$-\frac{3}{2}\sum \frac{L_v'(1)}{L_v(1)} \sum \log|n_1 n_3|_v + \frac{3}{2}\sum_v \frac{L_v'(1)}{L_v(1)} \sum_{w \neq v} \frac{L_w'(1)}{L_w(1)}$$

$$+ \sum[\tfrac{1}{2}(L_v(s)^{-1})''_{s=1} - (L_v(s)^{-1})'_{s=1}]L_v(1),$$

$$\frac{3}{2}\lambda_0/\lambda_{-1} \sum(\log|n_1 n_3|_v - 2L_v'(1)/L_v(1)) + (\lambda_0/\lambda_{-1})^2 + \lambda_1/\lambda_{-1}.$$

All sums are taken over any fixed finite set which includes all $v$ with $f_v \neq f_v^0$. This expression is the term (1) of Lemma 4.

2.5.5 The conjugacy class of the identity in $G$ yields

$$|G(F)Z(\mathbb{A})\backslash G(\mathbb{A})|f(1).$$

This is (3) of Lemma 4.

2.5.6 For the calculation of the remaining contributions put

$$u(x) = \begin{pmatrix} 1 & & x \\ & 1 & \\ & & 1 \end{pmatrix}.$$

Consider the terms in (3) which correspond to the conjugacy classes of $u(1)$ in $G$ and $1$ in $P_1$ and $P_2$. These yield

$$\int_{G_{u(1)}(F)Z(\mathbb{A})\backslash G(\mathbb{A})} f(g^{-1}u(1)g)\,dg$$

$$- \int_{P_2(F)Z(\mathbb{A})\backslash G(\mathbb{A})} \int_{N_2(\mathbb{A})} f(g^{-1}ng)\hat{\tau}_2(H_0(g)-T)dndg$$

$$- \int_{P_1(F)Z(\mathbb{A})\backslash G(\mathbb{A})} \int_{N_1(\mathbb{A})} f(g^{-1}ng)\hat{\tau}_1(H_0(g)-T)dndg$$

Applying the Poisson summation formula to each of the integrals over $N_1(\mathbb{A})$ and $N_2(\mathbb{A})$, which are viewed as Fourier transforms at 0, we see that up to terms whose sum tends to 0 as T goes to infinity this is

$$\int f(g^{-1}u(1)g)dg - \int \sum_{1 \neq n \in N_2(F)} f(g^{-1}ng)\hat{\tau}_2(H_0(g)-T)dg$$

$$- \int \sum_{1 \neq n \in N_1(F)} f(g^{-1}ng)\hat{\tau}_1(H_0(g)-T)dg$$

or

$$\int f(g^{-1}n(1)g)(1-\hat{\tau}_2-\hat{\tau}_1)(H_0(g)-T)dg \quad (g \text{ in } G_{u(1)}(F)Z(\mathbb{A})\backslash G(\mathbb{A}))$$

$$= \int_{P_0(F)Z_{\mathbb{A}}\backslash G_{\mathbb{A}}} f(g^{-1}n(x)g)(1-\hat{\tau}_2-\hat{\tau}_1)(H_0(g)-T)dg,$$

with a sum over $x$ in $F^\times$. This can also be expressed in this form

$$\int_{F^\times\backslash\mathbb{A}^\times}\int_{F^\times\backslash\mathbb{A}^\times} \sum f^K(u(xc/a))|c/a|^2[1-\chi(\log|a^2/c|>2t_2) - \chi(\log|a/c^2|>2t_2]d^\times a d^\times c$$

$$= \sum f^K(u(xc))|c|^2 d^\times c \quad (1-\chi(\log|a/c|>2t_1) - \chi(-\log|ac^2|>2t_2))d^\times a.$$

The inner integral here is $3 \log|c|$ if we ignore multiples of $t_1$ and $t_2$; we finally obtain

$$3\int_{F^\times\backslash\mathbb{A}^\times}\sum f^K(u(xc))|c|^2\log|c|d^\times c.$$

This is (2) of Lemma 4, whose proof is now complete.

### 2.6.1 Integration lemma

It remains to discuss the asymptotic behaviour of the $I(h,f)$, or rather the $J(h,f)$, as the regular split $h$ approaches the singular set. We saw that the limits were related to the values of the $I(h,f)$ and $J(h,f)$ on the singular set. But it does not suffice to find the limits of these distributions. Since we want to apply the Poisson summation formula to the non-smooth function $I(h,f)$ we have to check whether its asymptotic behaviour is not too bad and permits the application of this formula. The purpose of the following lemma is to show that the asymptotic behaviour which is discussed there affords the use of the formula. We shall show that the lemma applies in the case of $I(h,f)$.

Let $F$ be a local non-archimedean field.

LEMMA 8. *Let* $\psi$ *be a compactly supported function on* $F^\times \times F^\times$ *which is locally constant on the complement of* $\{(a,a);\ a \in F^\times\}$, $1 \times F^\times$ *and* $F^\times \times 1$. *Suppose that its asymptotic behaviour at* $(1,y)$ $(y \neq 1)$ *has the form*

$$b_0 + \sum b_{ij}|x-1|^i(\log|x-1|)^j \qquad (i \geq 1,\ j \geq 0),$$

*where the sum is finite and* $b_0$, $b_{ij}$ *are independent of* $x$. *Its asymptotic behaviour at* $(x,1)$ $(x \neq 1)$, *and at* $(x,x)$ $(x \neq 1)$, *is of the same form, with* y *replacing* x, *and* y-x *replacing* x-1, *respectively. Its asymptotic behaviour at* (1,1) *has the form*

$$(c_0 + \sum c_{ij}|x-1|^i(\log|x-1|)^j)(d_0 + \sum d_{ij}|y-1|^i(\log|y-1|)^j)$$

$$(e_0 + \sum e_{ij}|x-y|^i(\log|x-y|)^j),$$

*where the sums* (*over* $i \geq 1$, $j \geq 0$) *are finite and the constants are independent of* x *and* y. *Then its Fourier transform is absolutely integrable on the group of characters of* $F^\times \times F^\times$.

*Proof*. The Fourier transform $\hat{\psi}(\chi_1,\chi_2)$ of $\psi$ at a character $(\chi_1,\chi_2)$ of $F^\times \times F^\times$ is

$$\int_{F^\times \times F^\times} \psi(x,y)\chi_1(x)\chi_2(y)d^\times x d^\times y.$$

The group of characters of $F^\times$ is isomorphic to the direct product of the group of characters of the group of units in $F^\times$ and the group of unramified characters (which is a compact group, isomorphic to the unit circle). We denote by $p$ the maximal ideal in the ring of integers in $F^\times$, and say that $\chi$ has conductor $m(>0)$ if $\chi$ is trivial on $1+p^m$ but not on $1+p^{m-1}$. Such $\chi$ is denoted by $\chi_m$. Note that for each s in $p$ and $k(1 \leq k < m)$ we have

$$\sum_{x \text{ in } p^k/p^m} \chi_m(1+s+x) = 0 \qquad (*)$$

This follows from the existence of t in $p^{m-1}$ with $\chi_m(1+t) \neq 1$ and the identity

$$\chi_m(1+t)\sum_x \chi_m(1+s+x) = \sum \chi_m(1+s+x+t) = \sum \chi_m(1+s+x).$$

Suppose that $\psi$ is locally constant at $(1,1)$. We may assume that it is the characteristic function of some small neighborhood of $(1,1)$, of the form $\{1 + p^{k_1}\} \times \{1 + p^{k_2}\}$ $(k_1,k_2 > 0)$ In this case the integral over $(\chi_1,\chi_2)$ of $\psi(\chi_1,\chi_2)$ is reduced, by virtue of (*), to a finite sum (over characters of the group of units) of integrals (over the [compact] group of unramified characters) of a constant. This establishes the claim of the lemma if all $b_{ij}$, $c_{ij}$, . . . are 0. This also shows that the lemma is valid for functions which are everywhere locally constant, and so it remains to consider only functions which are supported on a small compact neighborhood intersecting $1 \times F^\times$, $1 \times F^\times$, or $\{(a,a);\ a \text{ in } F^\times\}$.

2.6.2 Suppose that $\psi$ is supported on a small neighborhood of $(1,y')$, where $y' \neq 1$. Since $\psi$ is locally constant in $y$ we can assume that $\psi$ is a product of a characteristic function of a compact set in $y$ and a compactly supported function of $x$ which is locally constant on the complement of 1 and with the above asymptotic behaviour at $x = 1$. We have seen that the Fourier transform of the first function is integrable. To study the second, which we denote by $\psi(x)$, it suffices to consider separately each summand from the sum describing the asymptotic behaviour. Since we may assume that $\psi(x)$ is supported on $1 + p$ it remains to prove the convergence, for each $i \geq 1$, $j \geq 0$, of the sum over all characters $\chi = \chi_m$ $(m \geq 1)$ of the group $1+p$ of the absolute values of

$$\int_{1+p} |x-1|^i(\log|x-1|)^j\chi_m(x)d^\times x = \int_p |x|^i(\log|x|)^j\chi_m(1+x)dx$$

$$= \sum_{x \in \mathfrak{p}/\mathfrak{p}^m} \chi_m(1+x) \int_{\mathfrak{p}^m} |x+s|^i (\log|x+s|)^j ds$$

$$= \sum_x \chi_m(1+x)|x|^i(\log|x|)^j \int_{\mathfrak{p}^m} ds + \int_{\mathfrak{p}^m} |s|^i(\log|s|)^j ds$$

$$- |\tilde{\omega}|^{im}(\log|\tilde{\omega}^m|)^j \int_{\mathfrak{p}^m} ds.$$

In the last sum (over $x$ in $\mathfrak{p}/\mathfrak{p}^m$) we take the representative $x = \tilde{\omega}^m$ for the class of $\mathfrak{p}^m$. We have the following identity:

$$\sum_{x \in \mathfrak{p}^k/\mathfrak{p}^m} \chi_m(1-x)|x|^i(\log|x|)^j$$

$$= \sum_{x \in \mathfrak{p}^{k+1}/\mathfrak{p}^m} \chi_m(1+x)(|x|^i(\log|x|)^j - |\tilde{\omega}|^{ik}(\log|\tilde{\omega}^k|)^j \qquad (1 \leq k < m)$$

$$= \sum_{x \in \mathfrak{p}^{k+1}/\mathfrak{p}^m} \chi_m(1+x)|x|^i(\log|x|)^j, \qquad (1 \leq k \leq m-2)$$

which is a consequence of (*). Our sum is the term with $k = 1$, and by induction we obtain the value at $m - 1$. We substitute the second line of this calculation with $k = m - 1$ in our sum and obtain

$$\int_{\mathfrak{p}^m} |s|^i(\log|s|)^j ds - |\tilde{\omega}|^{i(m-1)}(\log|\tilde{\omega}^{m-1}|)^j.$$

The absolute value of this expression is to be summed over the $\chi_m$. But the index of $\mathfrak{p}^m$ in the units of $F^\times$ is at most $q^m$, where $q = |\tilde{\omega}|^{-1}$. Hence the number of the $\chi_m$ is at most $q^m$ for each $m \geq 1$. Since $\sum_m m^j q^{-im}$ is finite for all $i \geq 1$, $j \geq 0$ our claim follows.

2.6.3 It remains to deal only with a function $\psi(x,y)$ with small support about (1,1). We may consider each term in the asymptotic behaviour separately. If at most two of $|x-1|, |y-1|, |x-y|$ occur non-trivially then we recover a previous case since on changing variables we see that the function is a product of two functions in a single variable. So we have to consider the case where $i_1$, $i_2$ and $i_3$ are all positive, and then prove the absolute convergence of the sum over $\chi_{1m}, \chi_{2n}$ whose general term is

$$\int_{1+p}\int_{1+p} |x-1|^{i_1}(\log|x-1|)^{j_1}|y-1|^{i_2}(\log|y-1|)^{j_2}|x-y|^{i_3}$$

$$(\log|x-y|)^{j_3}\chi_{1m}(x)\chi_{2m}(y)d^\times x d^\times y$$

$$= \int_p\int_p |x|^{i_1}(\log|x|)^{j_1}|y|^{i_2}(\log|y|)^{j_2}|x-y|^{i_3}(\log|x-y|)^{j_3}$$

$$\chi_{1m}(1+x)\chi_{2n}(1+y)dxdy$$

$$= \sum \chi_{1m}(1+s)\chi_{2n}(1+t)\int_{p^m}\int_{p^n} |s+x|^{i_1}(\log|s+x|)^{j_1}|t+y|^{i_2}$$

$$(\log|t+y|)^{j_2}|s-t+x-y|^{i_3}(\log|s-t+x-y|)^{j_3}dxdy.$$

Here the sum is taken over $s$ in $p/p^m$ and $t$ in $p/p^n$. We are going to rewrite this sum in terms of a finite number of partial sums and deal with each of these separately. We may assume that $m \leq n$, and for convenience introduce the notation $A_m(i,j)$ for $|\tilde{\omega}|^{im}m^j$. Since the number of $(\chi_{1m}, \chi_{2n})$ is bounded by $q^{m+n}$ each partial sum will be multiplied by this number. The result is a general term in a sum over $m$ and $n$ whose absolute convergence will be proved. The classes of $s$ in $p^m$ and $t$ in $p^n$ will be denoted by $\{0\}$.

We first consider the term with $s = \{0\}$ and $t = \{0\}$. Its product by $q^{m+n}$ is bounded by $A_m(i_1,j_1)A_n(i_2,j_2)$ which is a general term in a convergent sum. Next we consider the partial sum over $s = \{0\}$ and $t$ in $p/p^m$ with $t \neq \{0\}$. This sum is equal to

$$\sum_t \chi_{2n}(1+t)|t|^{i_2}(\log|t|)^{j_2}\int_{p^m}|x|^{i_1}(\log|x|)^{j_1}\int_{p^n}|x-t-y|^{i_3}(\log|x-t-y|)^{j_3}dydx.$$

If $m < n$ our inductive argument shows that the subsum here over $|t| > |\tilde{\omega}|^m$ is 0. The subsum over $|t| \leq |\tilde{\omega}^m|$ is bounded by a scalar multiple of $q^{-m-n}A_m(i_1,j_1)A_n(i_2,j_2)$. If $m = n$ then $|t| > |\tilde{\omega}^m|$ and we obtain the same bound (in fact a stronger one, where $(i_2,j_2)$ is replaced by $(i_2+i_3,j_2+j_3)$). The required convergence follows. Similarly the partial sum over $t = \{0\}$ and $s \neq \{0\}$, multiplied by $q^{m+n}$, is bounded by $A_m(i_1+i_3,j_1+j_3)A_n(i_2,j_2)$, and the required convergence of the resulting sum over $m,n$ follows at once.

There are two more partial sums to be considered. The first is over the $s$ with $s \neq \{0\}$ and the $t$ with $|s-t| \leq |\tilde{\omega}^m|$. In this case $|s| = |t|$ and $|t| > |\tilde{\omega}^m|$, and the sum becomes

$$\sum_t \chi_{1m}\chi_{2n}(1+t)|t|^{i_1+i_2}(\log|t|)^{j_1+j_2}\int_{p^m}\int_{p^n}|x-y|^{i_3}(\log|x-y|)^{j_3}dxdy$$

($t$ in $p/p^n$ but not in $p^m/p^n$). If $m < n$ then $\chi_{1m}\chi_{2n}$ has conductor $n$ and our inductive argument shows that the sum over $t$ is 0. If $m = n$ the character may have conductor less then $n$. But the absolute convergence of the sum over $t$ is clear since it is bounded by a scalar times a sum over $k(\leq m)$ of $A_k(i_1+i_2,j_1+j_2)$. The integrals on the right, multiplied by $q^{m+n}$, are bounded by a scalar times $A_m(i_3,j_3)$; the sum of these over $m$ is convergent, as required.

The last partial sum, multiplied by $q^{m+n}$, is of the form

$$\sum_s \chi_{1m}(1+s)|s|^{i_1}(\log|s|)^{j_1}\sum_t \chi_{2n}(1+t)|t|^{i_2}(\log|t|)^{j_2}|s-t|^{i_3}(\log|s-t|)^{j_3},$$

and the sums are taken over all $s$ (in $\mathfrak{p}/\mathfrak{p}^m$) and $t$ (in $\mathfrak{p}/\mathfrak{p}^n$) with $s \neq \{0\}$, $t \neq \{0\}$ and $|s-t| > |\tilde{\omega}^m|$. The subsum of the inner sum over $t$ with $|t| > |s|$ is the difference of a sum over all $t$ and a sum over $|t| \leq |s|$. Since $|s| > |\tilde{\omega}^m|$ these are equal by our inductive argument. If we considered only $t$ with $|t| < |s|$ then $|s-t| = |s|$ and we obtain a product of two sums, the first is bounded by $A_m(i_1+i_3, j_1+j_3)$ and the second by $A_n(i_2, j_2)$. So it remains to deal with the $t$ with $|t| = |s|$. We have the sum

$$\sum_{s \neq \{0\}} \chi_{1m}(1+s)|s|^{i_1+i_2}(\log|s|)^{j_1+j_2}\sum_t |s-t|^{i_3}(\log|s-t|)^{j_3}\chi_{2n}(1+t)$$

where $t$ satisfies $|t-s| > |\tilde{\omega}^m|$ and $|t| = |s|$. We may replace $s$ by $s + r$, sum over $r$ in $\mathfrak{p}^m/\mathfrak{p}^n$ and divide by the number $N$ of such $r$, without changing the expression under consideration. In the inner sum, over $t$, we replace $t$ by $t + s + r$ and obtain

$$\sum_s \chi_{1m}(1+s)|s|^{i_1+i_2}(\log|s|)^{j_1+j_2}\sum_t |t|^{i_3}(\log|t|)^{j_3}[\sum_{r \text{ in } \mathfrak{p}^m/\mathfrak{p}^n} \chi_{2n}(1+s+t+r)/N].$$

The sum over $r$ is $0$ if $m < n$ by virtue of (*). If $m = n$ it is equal to $1$, the sum over $t$ is convergent (as usual), and the sum over $s$ is bounded by a constant times $A_m(i_1+i_2, j_1+j_2)$. But the sum of these over $m$ is convergent, as required.

2.7.1 Asymptotic behaviour

The absolute convergence of the integral of the Fourier transform of $J(h,f_v)$ and $J_1(h,f_v)$ over the group of (unitary) characters of $A_0(F_v)$ modulo $N\,Z(E_v)$ is an elaborate excercise. For the archimedean places, in the simplest case one is reduced to a calculation involving $\int f(x,n)\log(|x|^2+|n|^2)dn$ ($n$ in $\mathbb{R}$ or $\mathbb{C}$) for a compactly supported smooth $f$, which was done in [12], §9. The generalization to functions in three (rather than one) variables, and quadratic weight factor, will be left for an amateur of archimedean places.

In the non-archimedean case we have to show that the asymptotic behaviour of $J(h,f_v)$ and $J_1(h,f_v)$ is of the type described by Lemma 8. The behaviour of $J_1$ is simpler than that of $J$; hence it suffices to consider only $J$. We have the local integral

$$\int f(hn)v(n)dn, \qquad v(n) = 2\log A \log B - (\log A/D)^2 - (\log B/C)^2,$$

where $f$ is compactly supported and locally constant, and

$$A = \|(y(z,n_3),\ n_1n_3-n_2z)\|, \qquad B = \|(y(x,n_1a/b),\ n_1n_3-n_2x)\|, \qquad C = \|(x,n_1a/b)\|,$$

$$D = \|(z,n_3)\|, \qquad x = 1-a/b, \qquad y = 1-c/a, \qquad z = 1-c/b.$$

We used the notation $\|(u(s,t),v)\|$ for $\|(us,ut,v)\|$.

The first case is when $h = (a,b,c) \longrightarrow (1,e,1)$. For $h$ sufficiently close to the limit $C$ and $D$ are independent of $h$, as well as $f(hn)$, and we shall content ourselves with a consideration of $\int f(hn)(\log A)^2dn$, noting that $z$ is now a constant. Since $f$ can be written as a sum $f_1+f_2$ with $f_1$ supported on the $n$ with $|n_3| \le |z|$ and $f_2$ on $|n_3| > |z|$, we shall consider the case where $f = f_1$.

The case where $f = f_2$ is similar and hence will be omitted. A change of variable $n_2 \longrightarrow (n_2+n_1n_3)/z$ reduces us to the study of

$$\int f(hn)(\log\|(y,n_2)\|)^2 dn = \int f(hn)(\log|n_2|)^2 dn + \int_{|n_2|<|y|} f(hn)[(\log|y|)^2-(\log|n_2|)^2]dn.$$

The first integral on the right is constant in $h$. In the second, since $f$ is locally constant $f(hn)$ is independent of $n_2$ if $|y|$ is sufficiently small; thus we replace $n_2$ by $0$ in $f(hn)$. The change $n_2 \longrightarrow n_2y$ of variables shows now that the second integral is the product of $|y|$ and a linear function of $\log|y|$.

2.7.2 <u>At $h = 1$</u>

It remains to look at the case when $h \longrightarrow 1$. Although the discussion of the integral of $f(hn)$ weighted by $\log A \log B$ is the longest, the techniques will be sufficiently exposed by the discussion of $\int f(hn)(\log A)^2 dn$, the only case to be described here. This last integral is

$$\int f(n)(\log|n_1n_3-n_2z|)^2 + \int_{|y|\,\|(z,n_3)\|>|n_1n_3-n_2z|} f(n)[(\log|y|\,\|(z,n_3)\|)^2 - (\log|n_1n_3-n_2z|)^2] \qquad (*)$$

The first is the sum of $\int f(n)(\log|n_1n_3|)^2$ (a constant), and

$$\int f(n)[(\log|n_1n_3-n_2z|)^2-\log|n_1n_3|)^2] \qquad \text{on} \qquad |n_1n_3| \le |n_2z|.$$

On the domain $|n_1| \le |n_3|$ we may replace $n_1$ by $0$ in $f(n)$ since $f$ is locally constant, change $n_1$ to $n_1z$, and consider the integral over $n_1$ with $|n_1n_3| \le |n_2|, |n_1z| \le |n_3|$, of $(\log|n_1n_3-n_2|\,|z|)^2 - (\log|n_1n_3z|)^2$. On $|n_2/n_3| \le |n_3/z|$ we have $|n_1| \le |n_2/n_3|$; multiplying $n_1$ by $n_2/n_3$ (for example) it is clear that the integral over $|n_1| \leq 1$ is 0. On $|n_2/n_3| > |n_3/z|$ we set $n_3 = 0$ in $f(n)$, and we have $|n_1| \leq |n_3/z|$; multiplying $n_1$ by $n_3/z$, and then $n_3$ by $z^{\frac{1}{2}}$ we are left with the integral over $|n_3^2| < |n_2|, |n_1| \leq 1$, of $|zn_3|$ multiplied by $f(n)(n_1=n_3=0)$ and $(\log|n_2z|)^2 - (\log|n_1n_3^2|)^2$, which is the product of $|y|$ and a quadratic polynomial in $\log|z|$. We omit the analogous discussion for the domain $|n_3| < |n_1|$.

The second integral in (*) is the sum of two integrals, $I_1$ and $I_2$. The first is the integral over $|n_3| < |z|, |n_2z-n_1n_3| < |yz|$, of $f(n)$ multiplied by $(\log|yz|)^2$ minus $(\log|n_1n_3-n_2z|)^2$. We may replace $n_3$ by 0 in $f(n)$, change $n_3$ to $n_3z$, and obtain the integral over $|n_3| < 1$, $|n_2-n_1n_3| < |y|$, of $|z|f(n)$ weighted by $(\log|yz|)^2$ minus $(\log|n_1n_3-n_2||z|)^2$. But now the change of $n_2$ to $n_2-n_1n_3$ shows that $n_2$ can be replaced by $n_1n_3$ in $f(n)$, and the change $n_2 \longrightarrow n_2y$ leaves us with the integral over $|n_3| < 1$, $|n_2| < 1$, of $|yz|f(n)$ times $(\log|yz|)^2$ minus $(\log|n_2yz|)^2$. This is the product of $|yz|$ and a linear polynomial in $\log|y|$ and $\log|z|$.

In the remaining integral $I_2$ of $f(n)$ times $(\log|yn_3|)^2$ minus $(\log|n_1n_3-zn_2|)^2$ over $|n_3| \geq |z|$, $|n_2z-n_1n_3| \geq |yn_3|$, we replace $n_1$ by $n_1 - n_2z/n_3$. We obtain an integral over $|n_1| < |y|, |n_3| > |z|$, hence replace $n_1$ by 0 in $f(n)$ and change $n_1$ to $n_1y$. The result is

$$|y| \int f\begin{pmatrix}1 & 0 & n_2 \\ & 1 & n_3 \\ & & 1\end{pmatrix} \int_{|n_1|<1} [(\log|yn_3|)^2 - (\log|yn_1n_3|)^2] - |yz| \int f\begin{pmatrix}1 & 0 & n_2 \\ & 1 & 0 \\ & & 1\end{pmatrix} \int_{|n_1|<1, |n_3|\leq 1} [(\log|yzn_3|)^2-(\log|yzn_1n_3|)^2],$$

which has the required form to which Lemma 8 applies.

### 2.7.3 Division algebras

The trace formula for GL(3), a major part of which was described in Lemma 7, simplifies considerably if for two distinct places $v$ the component $f_v$ of $f$ is cuspidal. We shall terminate this chapter by recording this statement, although the remaining term $\sum I_\chi(f)$ of the trace formula will be described only in chapter 4.

A smooth function $f_v$ is said to be cuspidal if for any $g,g'$ in $G(F_v)$ the integral $\int f_v(gng')dn$, over $n$ in any unipotent radical $N_i(F_v)$, is 0. For example, if $\pi_v$ is a square-integrable representation, consider its coefficient $f_v(g)=d(\pi_v)(\pi_v(g)u,\tilde{u})$ ($u$ and $\tilde{u}$ denote vectors in the space of $\pi_v$ and its contragrediant with $(u,\tilde{u}) = 1$). The value of the orbital integral of $f_v$ at elliptic regular $h$ is the character $\chi_{\pi_v}(h)$ of $\pi_v$ at $h$, and it is 0 at any other regular $h$. Such $f_v$ is cuspidal and in fact any function with equal orbital integrals is cuspidal. By the classification of orbital integrals, $f_v$ may be assumed to be compactly supported modulo the centre.

COROLLARY 9. *Suppose that for two distinct places* $v$ *the component* $f_v$ *of* $f = \otimes f_v$ *is cuspidal. Then the trace of the operator* $r(f)$ *is given by*

$$\operatorname{tr} r(f) = \sum_h |Z(\mathbb{A})G_h(F)\backslash G_h(\mathbb{A})| \int_{G_h(\mathbb{A})\backslash G(\mathbb{A})} f(g^{-1}hg)dg,$$

*where the sum is taken over conjugacy classes of the identity and of the regular elliptic elements* $h$ *in* $G(F)$ *modulo* $Z(F)$.

This lemma is a major step in the proof of the correspondence from the set of automorphic representations of the multiplicative group of a division algebra of dimension 9 over $F$ to that of $G = GL(3)$, and was assumed without proof in [4]. The terms in the sum are those described by (3) of Lemma 4 and by Lemma 1, where $Z_E(\mathbb{A})$ was replaced by $Z(\mathbb{A})$ (as in [4]).

The vanishing of (2), (3), (5) of Lemma 7 follows at once from the condition on $f_v$ for the two chosen places $v$. On expressing the second displayed line of (1), Lemma 3, and second, third lines of (1), Lemma 4, in terms of local orbital integrals, it is clear that they vanish as well. The same is true for (2) of Lemma 4, since $\int f_v^K(hu(c))|c|dc$ ($u(c)$ as in 2.6.6) is a scalar multiple of the orbital integral of $f_v$ at $hu(1)$. This orbital integral is a scalar multiple of $f_{vN_1}^K(h)$. The vanishing of the sum $\sum_\chi I_\chi(f)$ of (2) follows from its explicit description in chapter 4 below and the vanishing of $\mathrm{tr}\,\pi_v(f_v)$ for any induced representation $\pi$ if $f_v$ is cuspidal.

## §3. THE TWISTED TRACE FORMULA

### 3.1.1 Introduction

Let $\omega_E$ be the character $z \longmapsto \omega(Nz)$ of $Z(\mathbb{A}_E)$ and signify by $L^2(\omega_E)$ the space of measurable functions $\psi$ on $G(E)\backslash G(\mathbb{A}_E)$ with

$$\psi(zg) = \omega_E(z)\psi(g) \qquad (z \text{ in } Z(\mathbb{A}_E),\ g \text{ in } G(E)\backslash G(\mathbb{A}_E))$$

and

$$\int_{Z(\mathbb{A}_E)G(E)\backslash G(\mathbb{A}_E)} |\psi(g)|^2\, dg < \infty.$$

The group $G(\mathbb{A}_E)$ acts by right translations and $\mathcal{G}$ acts by $\varepsilon: \psi \longrightarrow {}^{\varepsilon}\psi$ where ${}^{\varepsilon}\psi(g) = \psi(h^{\varepsilon})$ ($\varepsilon$ in $\mathcal{G}$ ). The semi-direct product acts by $r((\varepsilon,g))\psi(g) = \psi(h^{\varepsilon}g)$. $L^2(\omega_E)$ decomposes as the direct sum of the invariant subspace $L_0^2(\omega_E)$ of square-integrable cusp forms and its orthogonal complement $L_c^2(\omega_E)$ which is also invariant.

Let $\phi = \otimes\phi_v$ be a function on $G(\mathbb{A}_E)$ such that for all places $v$ of $E$ the component $\phi_v$ is smooth (namely $K_v$-finite and highly differentiable in the archimedean case), compactly supported modulo $Z(E_v)$ and transforming under $Z(E_v)$ by $\omega_{E_v}^{-1}$. For almost all non-archimedean $v$ suppose that $\phi_v$ is $\phi_v^0$, the function which obtains the value $0$ unless $g_v = z_v k_v$ ($z_v$ in $Z(E_v)$, $k$ in $K_v$), where it is the quotient of $\omega_{E_v}^{-1}(z_v)$ by the measure of $K_v$.

Let $\sigma$ be a fixed non-trivial element in $\mathcal{G}$. Define the function $\phi'$ on $\mathcal{G} \times G(\mathbb{A}_E)$, which is supported on $\sigma \times G(\mathbb{A}_E)$, by $\phi'((\sigma,g)) = \phi(g)$. Consider the operator

$$r(\phi') = r(\sigma)r(\phi) = \int_{Z(\mathbb{A}_E)\backslash G(\mathbb{A}_E)} \phi(g)r((\sigma,g))dg$$

on $L_0^2(\omega_E)$. This is an integral operator of trace class with a kernal $K_0(h,g)$, described by the difference of $\sum\phi(h^{-\sigma}\gamma g)(\gamma \text{ in } Z(E)\backslash G(E))$ and an expression defined by means of Eisenstein series. The trace formula is an explicit expression for the integral of $K_0(h,g)$ over $h = g$ in $G(E)Z(\mathbb{A}_E)\backslash G(\mathbb{A}_E)$. In the case of $r(\phi')$ it is called the twisted trace formula. The explicit expression of use for us will be recorded here. The analogues of [1,2,3] will be written out elswhere. Note, however, that the effect of the twisting here can be easily traced through [1,2,3]. This is in contrast to the study of the outerly twisted trace formula in [5].

To describe the twisted trace formula recall (1.1.2) that the "norm" map

$$\gamma \longrightarrow N\gamma = \gamma^{\sigma^{\ell-1}} \cdots \gamma^{\sigma}\gamma \quad (\gamma \text{ in } G(E))$$

induces an injection from the set of $\sigma$-conjugacy classes in $G(E)$ to a set of conjugacy classes in $G(E)$ which intersect $G(F)$. Indeed in the sequel we shall denote by $N\gamma$ any element in $G(F)$ in the $G(E)$-conjugacy class of the image of $\gamma$ by $N$. We say that the elements $\gamma,\delta$ of $G(E)$ are $\sigma$-equivalent if the elements $N\gamma, N\delta$ of $G(F)$ are equivalent, namely if the semi-simple parts of $N\gamma$ and $N\delta$ are conjugate.

The twisted trace formula is the difference between two absolutely convergent sums:

$$\operatorname{tr} r(\phi') = \sum J_o(\phi) - \sum J_\chi(\phi) \tag{1}$$

The first sum is taken over $\sigma$-equivalence classes in $G(E)$ modulo $Z(E)$ or, what is the same, equivalence classes in the image of $N$ in $G(F)$ modulo $NZ(E)$.

A distribution $D$ is said to be $\sigma$-invariant if it obtains the same value at $\phi$ and at the function $g \longrightarrow \phi(h^{-\sigma}gh)$ for any $h$ in $G(\mathbb{A}_E)$. The distributions occurring in the above formula are not $\sigma$-invariant, but it is possible to rewrite it as a difference of two absolutely convergent sums,

$$\mathrm{tr}\, r(\phi') = \sum I_{\mathcal{O}}(\phi) - \sum I_{\chi}(\phi), \tag{2}$$

where the $I$ are $\sigma$-invariant. The $I_{\mathcal{O}}$ will be described after the description of the $J_{\mathcal{O}}$ is complete.

3.1.2 The twisted distribution $J_{\mathcal{O}}$

In the notations of chapter 2, $J_{\mathcal{O}}(\phi)$ is given by the constant term in $t_i = \frac{3}{2} \langle T, \mu_i \rangle$ $(i=1,2)$ of

$$\int_{G(E)Z(\mathbb{A}_E)\backslash G(\mathbb{A}_E)} \sum_P (-1)^{|A/Z|} \sum_{\delta} \hat{\tau}_P(H_0^E(\delta g)-T)$$

$$\sum_{\gamma} \sum_{\eta} \int_{N^{\sigma}_{\gamma_s}(\mathbb{A}_E)} \phi((\delta g)^{-\sigma} \eta^{-\sigma} n\eta\delta g)\, dn\, dg. \tag{3}$$

Here $\delta$ is taken over $P(E)\backslash G(E)$, $\gamma$ over $\mathcal{O} \cap M(E)$ and $\eta$ over $N^{\sigma}_{\gamma_s}(E)\backslash N(E)$. We denote by $\gamma_s$ an element of $G(E)$ such that $N(\gamma_s)$ is the semi-simple part of $N\gamma$. $N^{\sigma}_{\gamma_s}$ denotes the $\sigma$-centralizer of $\gamma_s$ in $N$.

The element $\gamma$ is said to be $\sigma$-regular ($\sigma$-elliptic, etc.) if $N\gamma$ is regular (elliptic, etc.). If the class $\mathcal{O}$ contains a $\sigma$-regular ($\sigma$-elliptic, etc.) element then we shall say that $\mathcal{O}$ is $\sigma$-regular ($\sigma$-elliptic, etc.).

If $\mathcal{O}$ contains a $\sigma$-regular element $\gamma$ and $M$ denotes the minimal group among $G$, $M_1$, $M_0$ such that $M(E)$ intersects $\mathcal{O}$ non-trivially, then $J_{\mathcal{O}}(\phi)$ has the explicit form

$$J_o(\phi) = \ell^{-q}|Z(\mathbb{A}_E)A(\mathbb{A})G_\gamma^\sigma(E)\backslash Z(\mathbb{A}_E)G_\gamma^\sigma(\mathbb{A}_E)|\int_{Z(\mathbb{A}_E)G_\gamma^\sigma(\mathbb{A}_E)\backslash G(\mathbb{A}_E)} \phi(g^{-\sigma}\gamma g)v_M^E(g)dg.$$

This is a twisted orbital integral which is obtained analogously to the $J_o(f)$ of chapter 2. It is weighted by a factor defined with respect to $E$, namely by the factor $v_M(g)$ of 2.2.1 in whose definition the function $H_i$ is replaced by $H_i^E$, where $H_i^E$ is the H-function defined with respect to the field $E$.

The coefficient can be simplified, as we shall do below, since

$$Z(\mathbb{A}_E)A(\mathbb{A})G_\gamma^\sigma(E)\backslash Z(\mathbb{A}_E)G_\gamma^\sigma(\mathbb{A}_E) \quad \text{and} \quad A(\mathbb{A})G_\gamma^\sigma(E)\backslash G_\gamma^\sigma(\mathbb{A}_E)$$

are isomorphic. The coefficient $\ell^{-q}$ appears since the functions $\hat{\tau}_P$ are defined with respect to the valuation on $E$.

### 3.1.3 Elliptic terms

The first part of the sum $\sum_o J_o(f)$ to be described is the following.

LEMMA 1. *The subsum over the classes $o$ containing $\gamma$ such that $N\gamma$ is either elliptic or a scalar in $F^\times$ but not in $NE^\times$ of the $J(N\gamma,\phi) = J_o(\phi)$ in (1) is equal to the sum over such $\sigma$-conjugacy classes $\{\gamma\}$, of*

$$\varepsilon(\gamma)|Z(\mathbb{A})G_\gamma^\sigma(E)\backslash G_\gamma^\sigma(\mathbb{A}_E)|\int_{Z(\mathbb{A}_E)G_\gamma^\sigma(\mathbb{A}_E)\backslash G(\mathbb{A}_E)} \phi(g^{-\sigma}\gamma g)dg.$$

*Here $\varepsilon(\gamma)$ is $\frac{1}{3}$ or $1$ depending on whether there do or do not exist $\delta$ in $G(E)$ and $z$ in $Z(E)$, but not in $Z(E)^{1-\sigma}$, such that $\delta^{-\sigma}\gamma\delta = z\gamma$.*

Note that $J(N\gamma,\phi)$ depends on $N\gamma$ but not on $\gamma$ itself.

If $o$ is $\sigma$-elliptic then $\{N\gamma;\ \gamma \text{ in } o\}$ is a conjugacy class and $o$ is therefore a $\sigma$-conjugacy class. A class $o$ which contains $\gamma$ such that $N\gamma$ lies in $F^\times$ but not in $NE^\times$ is also a $\sigma$-conjugacy class. Otherwise there is $\delta$ in such $o$ such that $N\delta$ is the product of $N\gamma$ and some non-trivial element of $N_0(F)$.

The centralizer of $N\delta$ in $G(E)$ is contained in $Z(E)N_0(E)$. Since $\delta$ centralizes $N\delta$ it follows that the semi-simple part of $N\delta$ lies in $NZ(E)$, which contradicts the assumption that $N\gamma$ does not lie in $NZ(E)$. The latter $\sigma$-conjugacy class exists only when $\ell = 3$, and we have $\varepsilon(\gamma) = 1$ by "Hilbert's theorem 90" for its elements.

The sum of the lemma is obtained by integrating over $G(E)Z(\mathbb{A}_E)\backslash G(\mathbb{A}_E)$ the sum $\sum \phi(g^{-\sigma}\gamma g)$ over our $\gamma$ in $G(E)$ mod $Z(E)$, which can also be expressed as a sum over $\sigma$-conjugacy classes of such $\gamma$, namely

$$\sum_{\gamma} \varepsilon(\gamma) \sum_{\delta \text{ in } Z(E)G_{\gamma}^{\sigma}(E)\backslash G(E)} \phi(g^{-\sigma}\delta^{-\sigma}\gamma\delta g).$$

3.1.4 Quadratic terms

Here we shall consider the classes $\mathcal{O}$ which contain $\gamma$ such that $N\gamma$ is either quadratic in $G(F)$ or exactly two of its eigenvalues are equal and lie in $F^{\times}$ but not in $NE^{\times}$. Such $\mathcal{O}$ and $\gamma$ will be called quadratic in the present subsection. We shall assume that $\gamma$ lies in $M_1(E)$ without any loss of generality.

As noted in 3.1.2 the distribution $J_{\mathcal{O}}(\phi)$ is obtained by integrating over $g$ in $G(E)Z(\mathbb{A}_E)\backslash G(\mathbb{A}_E)$ the sum over $\delta$ in $Z(E)G_{\gamma}^{\sigma}(E)\backslash G(E)$ of the product of $\phi(g^{-\sigma}\delta^{-\sigma}\gamma\delta g)$ and the expression (8.5) of [1]. The integral obtained by combining the integral and sum is expressed as two, an inner integral over $Z(\mathbb{A})\backslash A_1(\mathbb{A})$, and one over $Z(\mathbb{A}_E)A_1(\mathbb{A})G_{\gamma}^{\sigma}(E)\backslash G(\mathbb{A}_E)$. Taking $\phi(g^{-\sigma}\delta g)$ out, the inner integral becomes $\ell^{-1}v_1^E(g)$.

The local analogue of the integral which now defines $J_{\mathcal{O}}(\phi)$ is

$$J_{\mathcal{O}}(\phi_v) = \Delta_v(N\gamma)\int_{Z(E_v)G_{\gamma}^{\sigma}(E_v)\backslash G(E_v)} \phi_v(g^{-\sigma}\gamma g)\,\log\|(1,n_2,n_3)\|_v,$$

where $g = mnk$ according to the decomposition $M_1N_1(K\cap M_1\backslash K)$ of $G$. It will be shown below that in the case that the set $\{h_1,h_2,h_3\}$ of eigenvalues of $N\gamma$ is contained in $F^\times$(hence $h_1=h_2$), the function $J_o(\phi_v)$ has a singularity as $N\gamma \longrightarrow 1$. To remove it we introduce the corrected twisted orbital integral

$$J(N\gamma,\phi_v) = \Delta_v(N\gamma) \iiint_{Z(E_v)M^\sigma_{1\gamma}(E_v)\backslash M_1(E_v)} \phi_v(k^{-\sigma}n^{-\sigma}m^{-\sigma}\gamma mnk)\ \log\|(1,n_2,n_3)z\|_v dmdndk,$$

where $z = 1 - h_3/h_1$; cf. (2) of Lemma 2.3, and 2.4.4.

If the eigenvalues of $N\gamma$ do not lie in $F^\times$ we do not have to correct $J_o(\phi_v)$ for the purposes of the current work, and we put $J(N\gamma,\phi_v) = J_o(\phi_v)$. However if it becomes necessary to remove the singularity of $J_o(\phi_v)$ as $N\gamma \longrightarrow 1$, they can be corrected by means of the $z$ found in the proof of Lemma 3 below.

Let $F(N\gamma,\phi_w)$ be the twisted orbital integral (defined as $J$, but with the weight factor omitted). Note that $\lambda_{-1}$ is the discrepancy between the global and product of local measures on $M_1 \cap K\backslash K$. The product formula on $F^\times$ implies the following.

LEMMA 2. <u>The sum of</u> $J_o(\phi)$ <u>over the classes under consideration is equal to the sum over a set of representatives</u> $\gamma$ <u>in</u> $M_1(E)$ <u>for the</u> $\sigma$-<u>conjugacy classes of quadratic elements in</u> $A_1(E)\backslash M_1(E)$ <u>of</u>

$$\frac{3}{2}\lambda_{-1}\ell^{-1} \sum_a |A_1(\mathbb{A})G^\sigma_\gamma(E)\backslash G^\sigma_\gamma(\mathbb{A}_E)| \sum_v J(N(a\gamma),\phi_v) \prod_{w\neq v} F(N(a\gamma),\phi_w).$$

<u>The sum over</u> $a$ <u>extends over</u> $Z(E)A_1^{1-\sigma}(E)\backslash A_1(E)$.

Note that if $g^{-\sigma}\gamma g = z\gamma$ with $g$ in $G(E)$, $z$ in $Z(E)$, then $g^{-1}N\gamma g = (Nz)N\gamma$, so that $g$ lies in $M_1(E)$ and hence $z$ is in $Z(E)^{1-\sigma}$.

LEMMA 3. <u>The function</u> $J(N(a\gamma),\phi_v)$ <u>of</u> $a$ <u>in</u> $A_1(E_v)$ <u>is smooth and compactly supported modulo</u> $Z(E_v)$. <u>As a function of</u> $N\gamma$ <u>in</u> $A_1(F)$ <u>it extends continuously to the entire subgroup of</u> $A_1(F)$ <u>which consists of norms</u>.

For brevity we drop the index $v$ in the proof, assume that $\gamma = \mathrm{diag}\left(\begin{pmatrix} a & b \\ c & d \end{pmatrix}, e\right)$ with entries in $E$, and put $x = ad - bc$. Denote $a_1 = ae/x$, $b_1 = be/x$, $c_1 = ce/x$, $d_1 = de/x$, and consider the linear transformation

$$n = \begin{pmatrix} 1 & 0 & n \\ & 1 & m \\ & & 1 \end{pmatrix} \longmapsto n' = \gamma^{-1} n^{-\sigma} \gamma n \qquad \text{by } \begin{pmatrix} n \\ m \end{pmatrix} \longmapsto \begin{pmatrix} n - d_1 n^\sigma + b_1 m^\sigma \\ c_1 n^\sigma + m - a_1 m^\sigma \end{pmatrix},$$

over $F$. Since the vector space $\mathrm{Res}_{E/F} E$ obtained from $E$ by restricting scalars from $E$ to $F$ is isomorphic to the direct sum of $\ell$ copies of $E$, when expressed with respect to the basis $(n_1, m_1, n_2, m_2, \ldots, n_\ell, m_\ell)$ the transformation is represented by the matrix

$$\begin{pmatrix}
1 & 0 & -d_1 & b_1 & & & & \\
0 & 1 & c_1 & -a_1 & & & & \\
 & & 1 & 0 & -d_1^\sigma & b_1^\sigma & & \\
 & & & 1 & c_1^\sigma & -a_1^\sigma & & \\
\vdots & \vdots & & & \ddots & \ddots & & \\
 & & & & 1 & 0 & -d_1^{\sigma^{\ell-2}} & b_1^{\sigma^{\ell-2}} \\
 & & & & & 1 & c_1^{\sigma^{\ell-2}} & -a_1^{\sigma^{\ell-2}} \\
-d_1^{\sigma^{\ell-1}} & b_1^{\sigma^{\ell-1}} & & \cdots & & & 1 & 0 \\
c_1^{\sigma^{\ell-1}} & -a_1^{\sigma^{\ell-1}} & & \cdots & & & 0 & 1
\end{pmatrix}$$

When $\ell = 2$ the determinant is equal to

$$z = 1 - N(e/x)[aa^\sigma + dd^\sigma + bc^\sigma + cb^\sigma] + N(e^2/x)$$

$$= (Nx - (\mathrm{tr}\, N\begin{pmatrix} a & b \\ c & d \end{pmatrix}) Ne + Ne^2)/Nx.$$

This includes the case when exactly two of the eigenvalues of $N\gamma$ are equal, since then $\ell$ is necessarily equal to 2.

In fact in the last case if we take $\gamma$ to be defined by $a = d = 0$, $b = 1$, $c = f$, where $f$ is the eigenvalue with multiplicity two, it is clear that the inverse transformation is given by $n \longrightarrow z^{-1}(n + m^{\sigma}e/f)$, $m \longrightarrow z^{-1}(m + en^{\sigma})$. Noting that each entry in the weight factor is multiplied by $z$ we obtain the continuity of $J(N\gamma,\phi)$ on the subgroup of $A_1(F)$ which consists of norms.

The last expression for $z$ is in fact the determinant for all $\ell$. To see this when $N\gamma$ generates a quadratic torus $T$ which does not split over $E$, and in particular when $\ell > 2$, we note that $\gamma$ lies in $T(E)$ and hence it is $\sigma$-conjugate to some

$$\operatorname{diag}\left(\begin{pmatrix} a & b\theta \\ b & a \end{pmatrix}, e\right) = u^{-1}\delta u \quad \text{with} \quad \delta = \operatorname{diag}\,(a+b\sqrt{\theta}, a-b\sqrt{\theta}, e)$$

$$\text{and} \quad u = \operatorname{diag}\left(\begin{pmatrix} 1 & \sqrt{\theta} \\ -\frac{1}{2}/\sqrt{\theta} & \frac{1}{2} \end{pmatrix}, 1\right),$$

where $\theta$ lies in $F^{\times}$ and its square root generates $T$ over $F^{\times}$. But $u^{\sigma} = u$ so that $\gamma = m^{-\sigma}\delta m$ with $m$ in $M_1(E)$. Hence $n' = \gamma^{-1}n^{-\sigma}\gamma n$ is equal to $m^{-1}[\delta^{-1}(mnm^{-1})^{-\sigma}\delta(mnm^{-1})]m$. In other words, $n'$ is obtained from $n$ by applying the transformation $A^{-1}DA$ where $A$ maps $n$ to $mnm^{-1}$ and $D$ maps $n$ to $\delta^{-1}n^{-\sigma}\delta n$. Now if $\delta = \operatorname{diag}(\gamma_1, \gamma_2, e)$ and

$$n = \begin{pmatrix} 1 & 0 & n \\ & 1 & m \\ & & 1 \end{pmatrix}, \quad \text{then} \quad \delta^{-1}n^{-\sigma}\delta n = \begin{pmatrix} 1 & 0 & n-n^{\sigma}e/\gamma_1 \\ & 1 & m-m^{\sigma}e/\gamma_2 \\ & & 1 \end{pmatrix}.$$

But the determinant of the transformation $x \longrightarrow x - bx^{\sigma}$ ($b$ in $T(E)$) of $T(E)$ as a vector space over $T(F)$ is $1 - N_{T(E)/T(F)}b$ ([12], Lemma 4.5). It follows that $z = \det D = \det A^{-1}DA$ is equal to

$$z = (N(\gamma_1\gamma_2) - (N\gamma_1 + N\gamma_2)Ne + Ne^2)/N\gamma_1\gamma_2,$$

as required.

The transformation inverse to $n \longrightarrow n'$ is therefore given by

$$(n,m) \longmapsto (n^*,m^*) = (z^{-1}\sum(p_{1j}n^{\sigma^j}+p_{2j}m^{\sigma^j}),z^{-1}\sum(p_{3j}n^{\sigma^j}+p_{4j}m^{\sigma^j})),$$

where the sums are taken over $j(0 \le j \le \ell-1)$ and where $p_{ij}(1 \le i \le 4)$ are polynomials with integral coefficients in $ae/x$, $be/x$, $ce/x$, $de/x$, and their conjugates. Applying this inverse transformation and noting that

$$\Delta(N\gamma) = \left|\frac{N\gamma_1 - Ne}{N\gamma_1}\ \frac{N\gamma_2 - Ne}{N\gamma_2}\ \frac{N\gamma_1 - N\gamma_2}{Ne}\right|,$$

the integral $J(N\gamma,\phi)$ becomes

$$|(N\gamma_1 - N\gamma_2)/Ne| \iiint \phi(k^{-\sigma}m^{-\sigma}\gamma mnk)dmdk < -w\alpha, H_1^E\left(\left(\begin{matrix} 1 & & \\ n^* & 1 & \\ m^* & 0 & 1\end{matrix}\right)\right) >/<\alpha, w^{-1}\alpha_2> dn. \qquad (*)$$

The first assertion of the lemma follows now at once.

LEMMA 4. *Let* $v_0$ *be a fixed place of* $F$. *Then the sum over* $v$ *in Lemma* 2 *can be taken only over a finite set of places of* $F$. *This set may depend on* $\gamma$ *and (the support of)* $\phi_w$ *for* $w \ne v_0$, *but it is independent of* $\phi_{v_0}$.

The twisted orbital integrals $F(N\gamma,\phi_v)$ are described by (*), with the weight factor omitted. Since $\phi_v$ are compactly supported modulo $Z(E_v)$ there are $C_v \ge 1$ $(v \ne v_0)$ with $C_v = 1$ for all $v$ with $\phi_v = \phi_v^0$, with the following property. If $\Pi_v\phi_v(k^{-\sigma}m^{-\sigma}a\gamma mnk) \ne 0$ for some $n$ in $N_1(\mathbb{A})$, $k$ in $K(\mathbb{A})$, $m$ in $M_1(\mathbb{A})$, $a$ in $A_1(E)$, then $|N(b/c)|_v$ lies between $C_v$ and $C_v^{-1}$ for all $v \ne v_0$. Here $b$ and $c$ denote the eigenvalues of $a$, and $N$ is the norm form $E$ to $F$. The product formula on $F$ implies this for the remaining place $v_0$, with $C_{v_0} = \Pi C_v(v \ne v_0)$.

Hence $N(b/c)$ lies in a compact subgroup, and consequently the sum over $a$ in Lemma 2 can be taken over a finite set, independent of $\phi_{v_0}$. For each such $a$ we claim that $J(N(a\gamma),\phi_v) \neq 0$ only for $v$ in a finite set independent of $\phi_{v_0}$. Indeed, outside a finite set which depends only on $\gamma$ we have $|Nz|_v = |N(x/e^2)|_v = 1$ If $\phi_v = \phi_v^0$ then (the norms of) $a/e$, $b/e$, $c/e$, $d/e$, $n$ and $m$ (see proof of Lemma 3) are bounded by $1$ in the v-valuation, hence $n^*$, $m^*$ are v-integral, the weight factor in (*) vanishes and $J(N(a\gamma),\phi_v)$ is $0$, and the lemma follows.

### 3.2.1 Twisted correction for $GL(2,E)$

It remains to discuss the twisted weighted orbital integrals $J_{\mathcal{O}}(\phi)$ for the split classes $\mathcal{O}$ which contain a diagonal element. As the summation formula will be applied to their sum, they will have to be corrected so as to have the required properties, analogous to those of $J(h,f)$ in the non-twisted case.

Again we shall discuss first the case of the distribution $J_{1\mathcal{O}}(\phi)$ on $G(\mathbb{A}_E)$, where now $G = GL(2)$. If $\mathcal{O}$ is the class of $\gamma = \begin{pmatrix} \gamma_1 & \\ & \gamma_2 \end{pmatrix}$ and $N(\gamma_2/\gamma_1) \neq 1$, then $J_{1\mathcal{O}}(\phi)$ is given by the twisted orbital integral of $\phi$ weighted by

$$v^E(g) = 2\sum_v \log\|(1,n_v)\|_{E_v} \qquad (g = a\begin{pmatrix} 1 & n \\ 0 & 1 \end{pmatrix}k,\ a \text{ in } A(\mathbb{A}_E),\ n = (n_v) \text{ in } \mathbb{A}_E).$$

This is the same expression as in 2.3.1 with the function $H$ of 2.1.3 replaced by its analogue $H^E$ over $E$. The sum over $v$ ranges over all places of $E$; it is a finite sum for each $n$ in $\mathbb{A}_E$. Then

$$J_{1\mathcal{O}}(\phi) = \ell^{-1}\int_{Z(\mathbb{A}_E)G_\gamma^\sigma(\mathbb{A}_E)\backslash G(\mathbb{A}_E)} \phi(g^{-\sigma}\gamma g)v^E(g)dg$$

$$= 2\ell^{-1}\int_{N(\mathbb{A}_E)}\int \phi^K(n^{-\sigma}a^{-\sigma}\gamma an)\sum_v \log\|(1,n_v)\|_{E_v}\,da\,dn \quad (a = \begin{pmatrix} a & \\ & b \end{pmatrix} \text{ in } Z(\mathbb{A}_E)A(\mathbb{A})\backslash A(\mathbb{A}_E))$$

Here we put $\phi^K(g) = \int (k^{-\sigma}gk)dk$ (k in K). The analogous notation will be employed below in the local case, and in the context of GL(3,E).

It can be seen that as a function of $\gamma$ the function $J_{1o}(\phi)$ has a singularity at the singular set. Motivated by the non-twisted case we introduce the corrected twisted weighted orbital integral $J_1(N\gamma,\phi)$, which depends on $N\gamma$ but not on $\gamma$ itself, to be

$$J_1(N\gamma,\phi) = \frac{1}{\ell}\iint \phi^K(n^{-\sigma}a^{-\sigma}\gamma an) \sum_v \log\|(1,n_v)(1-N(\gamma_2/\gamma_1))\|_{E_v} \, da\, dn,$$

where

$$\|(x,y)z\| = \|(xz,yz)\|.$$

Since $1 - N(\gamma_2/\gamma_1)$ lies in $F^\times$ the product formula (on $E^\times$) implies that

$$\sum_\gamma J_1(N\gamma,\phi) = 2J_1(N\gamma,\phi) = J_{1o}(\phi);$$

here $\gamma$ is taken in the intersection of $\mathcal{O}$ and $Z(E)A^{1-\sigma}(E)\setminus A(E)$.

The local analogue $J_1(N\gamma,\phi_v)$ is defined for each place $v$ of $F$ by the product of $\Delta_v(N\gamma)$ and the local variant of the integral which defines $J_1(N\gamma,\phi)$. The sum is now taken over the places of $E$ which lie above $v$, and $\phi_v$ is a product of $\ell$ functions if $E \otimes_F F_v$ is not a field. Further, $F(N\gamma,\phi_v)$ is defined in the same way with the weight factor omitted. Hence when $\gamma$ is $\sigma$-regular we have the local expression

$$J_1(N\gamma,\phi) = \lambda_{-1}\ell^{-1} \sum_v J_1(N\gamma,\phi_v) \prod_{w\neq v} F(N\gamma,\phi_w). \tag{4}$$

LEMMA 5. <u>Let</u> $v_0$ <u>be a fixed place of</u> F. <u>Then the sum over the places</u> v <u>of</u> F <u>in</u> (4) <u>can be taken only over a finite set independent of</u> $\phi_{v_0}$ <u>and</u> $\gamma$; <u>this set includes all</u> v <u>with</u> $\phi_v \neq \phi_v^0$ <u>or</u> $|1 - N(\gamma_2/\gamma_1)|_v \neq 1$.

The standard argument shows that if $\phi^K(n^{-\sigma}a^{-\sigma}\gamma an) \neq 0$ for some $a$ and $n$ then $|N(\gamma_2/\gamma_1)|_v$ lies between $C_v$ and $C_v^{-1}$ for all $v \neq v_0$, where $C_v \geq 1$ depends only on $\phi_v$, and $C_v = 1$ if $\phi_v = \phi_v^0$. Hence $\phi^K$ is non-zero only if $\gamma$ lies in a finite set which is independent of $\phi_{v_0}$. Note that the inverse of the transformation $An = n - n^\sigma b$ is given by

$$A^{-1}n = (1 - Nb)^{-1}(n + \sum_{i=1}^{\ell-1} n^{\sigma^i} \sum_{j=0}^{i-1} b^{\sigma^i}) \qquad (n \text{ in } \mathbb{A}_E,\ Nb \neq 1).$$

For almost all $v$ both $|\gamma_2/\gamma_1|_{E_v}$ and $|1 - N(\gamma_2/\gamma_1)|_v$ are 1 for all of the $\gamma$ in the above set. Then $|An|_{E_v} \leq 1$ if and only if $|n|_{E_v} \leq 1$ $(b = \gamma_2/\gamma_1)$, and if an addition $\phi_v = \phi_v^0$ then the weight factor and $J_1(N\gamma,\phi_v)$ vanish, as required.

We shall now verify that $J_1(N\gamma,\phi)$ extends continously to the entire group $Z(\mathbb{A}_E)A^{1-\sigma}(\mathbb{A}_E)\backslash A(\mathbb{A}_E)$ as a function of $\gamma$. Suppose that $N\gamma$ is close to 1. Then we may assume that $\gamma$ lies in $A(\mathbb{A})$ and it is near 1. In the integral which defines $J_1(N\gamma,\phi)$ we change the variable $n$ so that $an$ becomes $na$, then replace $n$ by $n^F n$ where $n^F$ lies in $N(\mathbb{A})$ and $n$ in $N(\mathbb{A})\backslash N(\mathbb{A}_E)$, then change $na$ back to $an$, and finally change $n^F$ so that $(n^F)^{-1}\gamma n^F$ becomes $\gamma n^F$. The effect of the last transformation is to multiply $n^F$(in $\mathbb{A}$) by $(1-\gamma_2/\gamma_1)^{-1}$ (in $\mathbb{A}^\times$). Hence $J_1(N\gamma,\phi)$ is equal to

$$\ell^{-1}\iiint \phi^K(n^{-\sigma}a^{-\sigma}\gamma n^F an) \sum_v \log\|(1,n_v+\frac{b}{a}n_v^F/(1-\gamma_2/\gamma_1))(1-N(\gamma_2/\gamma_1))\|_{E_v} \; |b/a|_{E_v}^{1/\ell}.$$

The last expression of $J_1(N\gamma,\phi)$ shows that its limit as $\gamma_2/\gamma_1 \longrightarrow 1$ is the first summand in the following expression [12] for $J_{1o}(\phi)$, where $o$ is the class of the identity. Put $n_0 = \begin{pmatrix}1 & 1/\ell\\ 0 & 1\end{pmatrix}$.

$$J_{1o}(\phi) = \lambda_{-1}\ell^{-1} \sum_v L_v(1)^{-1} \iint \phi_v^K(n^{-\sigma}a^{-\sigma}n_0 an)|a/b|_{E_v}^{-1/\ell} \log|b/a|_{E_v} \, dadn$$

$$\prod_{w\neq v} L_w(1)^{-1} \iint \phi_v^K(n^{-\sigma}a^{-\sigma}n_0 an)|a/b|_{E_v}^{-1/\ell} \, dadn \qquad (a = \begin{pmatrix}a & 0\\ 0 & b\end{pmatrix})$$

$$+ (\lambda_0-\lambda_{-1} \sum_v L_v'(1)/L_v(1)) \prod_v L_v(1)^{-1} \iint \phi_v(n^{-\sigma}a^{-\sigma}n_0 an)|a/b|_{E_v}^{-1/\ell} \,dadn$$

$$+ |G(F)Z(\mathbb{A})\backslash G(\mathbb{A})| \int_{Z(\mathbb{A}_E)G(\mathbb{A})\backslash G(\mathbb{A}_E)} \phi(g^{-\sigma}g)dg$$

The sums are taken over any set which contains all $v$ where $\phi_v \neq \phi_v^0$. The double integrals are taken over $N(F_v)\backslash N(E_v)$ and $Z(E_v)\backslash A(E_v)$. That the first summand is the required limit follows on replacing $a$ by $a^F a$ ($a^F$ in $Z(E_v)\backslash A(F_v)$) and expressing $(a^F)^{-1}n_0 a^F$ as $n^F$.

As in the non-twisted case the corrected twisted integral can be defined by replacing $H^E(wg)$ by $H^E(wg) - \tilde{H}^E(w\rho)$ in the original definition of $v^E(g)$. $\tilde{H}^E$ is the analogue over $E$ of $\tilde{H}$ (over $F$), and $\rho$ is defined by $(N\gamma)^{-1}\rho^{-1}N\gamma\rho = \begin{pmatrix} 1 & 1 \\ 0 & 1 \end{pmatrix}$.

### 3.2.2 Twisted correction for $GL(3,E)$

The global distribution $J_o(\phi)$ for a regular twisted class $o$ which contains $\gamma$ in $A(E)$ is defined by

$$J_o(\phi) = \tfrac{3}{2}\ell^{-2} \iint \phi^K(n^{-\sigma}a^{-\sigma}\gamma an)v_0^E(n)dadn \quad (n \text{ in } N_0(\mathbb{A}_E),\ a \text{ in } A_0(\mathbb{A})Z(\mathbb{A}_E)\backslash A_0(\mathbb{A}_E)).$$

Here $v_0^E(n)$ is given by (4) of 2.3.2 where the valuation in A, B, C, D are taken with respect to $E$. As a function of $\gamma$ in $A(\mathbb{A}_E)$ this function is not regular near the singular set (of $N\gamma$ with several equal eigenvalues), and as in 3.2.1, motivated by the non-twisted case we multiply each of the entries of A, B, C, D by the same quantity as in 2.3.2, but with $h$ there replaced by $N\gamma$ here. Thus we can introduce the corrected twisted weighted orbital integral of $\gamma$ in $o \cap A(E)$ to be

$$J(N\gamma,\phi) = \tfrac{1}{4}\ell^{-2} \iint \phi^K(n^{-\sigma}a^{-\sigma}\gamma an)v_0^E \,dadn,$$

where the weight factor is again given by (4) of 2.3.2, and

$$A = \|(1, n_3, n_2 - n_1 n_3) yz\|_{\mathbb{A}_E}, \qquad B = \|(1, n_1, n_2) xy\|_{\mathbb{A}_E}, \qquad C = \|(1, n_1) x\|_{\mathbb{A}_E}$$

$$D = \|(1, n_3) z\|_{\mathbb{A}_E}, \quad x = 1 - N(\gamma/\gamma_2), \quad y = 1 - N(\gamma_3/\gamma_1), \quad z = 1 - N(\gamma_3/\gamma_2),$$

where $\gamma = \mathrm{diag}(\gamma_1, \gamma_2, \gamma_3)$.

The $J(N\gamma, \phi)$ depends on $N\gamma$, but not on $\gamma$ itself. Since $x, y, z$ lie in $F^\times$ the product formula (on $E^\times$) implies that

$$\sum_{N\gamma} J(N\gamma, \phi) = 6J(N\gamma, \phi) = J_{\mathcal{O}}(\phi),$$

where $\gamma$ lies in the intersection of $\mathcal{O}$ and $A_0^{1-\sigma}(E)Z(E)\backslash A_0(E)$; alternatively $N\gamma$ lies in the intersection of the image of $\mathcal{O}$ under $N$ with $NZ(E)\backslash NA_0(E)$.

If $v_0$ is a fixed place of $F$ and all components of $\phi$ except $\phi_{v_0}$ are fixed, the usual argument shows that $\phi^K(n^{-\sigma}a^{-\sigma}\gamma an)$ is non-zero only if $N\gamma$ lies in a finite set of $NZ(E)\backslash NA_0(E)$ which is independent of $\phi_{v_0}$. The argument of Lemma 5 applied to the matrix $n^{-\sigma}a^{-\sigma}\gamma an$ shows that the product over the places of $v$ which defines the volumes $A, B, C, D$ can be taken only over those places where $\phi_w \neq \phi_w^0$ or $|x|_w, |y|_w, |z|_w$ are not $1$ for any place $w$ of $E$ above $v$.

We have to verify the following.

LEMMA 6. <u>The corrected twisted weighted orbital integral</u> $J(N\gamma, \phi)$ <u>extends continuously as a function of</u> $\gamma$ <u>on the entire space</u> $Z(\mathbb{A}_E)A_0^{1-\sigma}(\mathbb{A}_E)\backslash A_0(\mathbb{A}_E)$.

To apply the summation formula we shall have to show that the asymptotic behaviour at the singular set is of the type of Lemma 8 in 2.6.1, locally. Here we shall begin with the proof of the lemma.

There are two cases to be considered in the proof of Lemma 6. The first is when $N\gamma$ approaches a diagonal element with exactly two equal eigenvalues. Since $\phi$ transforms under the centre by a character we may assume that the limit is of the form $\mathrm{diag}(1,N\alpha,1)$, and when $N\gamma$ is sufficiently close to the limit we may further assume that $\gamma \longrightarrow \mathrm{diag}(1,\alpha,1)$, and that $\gamma_1$ and $\gamma_3$ lie in $\mathbb{A}^\times$.

To find the limit we change $n$ so that $an$ is replaced by $na$ in $J(N\gamma,\phi)$, then write $n$ as $n^F n$ with $n^F$ in $(N_1 \cap N_2)(\mathbb{A})$ and $n$ in $(N_1 \cap N_2)(\mathbb{A})\backslash N_0(\mathbb{A}_E)$, and change $na$ back to $an$. Then $(n^F)^{-1}\gamma n^F$ is replaced by $\gamma n^F$ if the component $n^F$ in $\mathbb{A}$ of the matrix $n^F$ is multiplied by $(1-\gamma_3/\gamma_1)^{-1}$ (an element of $\mathbb{A}^\times$). We obtain $\frac{1}{4}\ell^{-2}$ times the integral over $n$ in $(N_1 \cap N_2)(\mathbb{A})\backslash N_0(\mathbb{A}_E)$, $a$ in $Z(\mathbb{A}_E)A_0(\mathbb{A})\backslash A_0(\mathbb{A}_E)$, $n^F$ in $(N_1 \cap N_2)(\mathbb{A})$, of the product of $\phi^K(n^{-\sigma}a^{-\sigma}\gamma n^F an)\|\frac{c}{a}\|^{1/\ell}_{\mathbb{A}_E}$ by the weight factor (4) of 2.3.2 where

$$A = \|(1,n_3,n_2-n_1n_3+n^F c/a(1-\gamma_3/\gamma_1))yz\|_{\mathbb{A}_E} \qquad a = \mathrm{diag}(a,b,c)$$

$$B = \|(1,n_1,n_2+n^F c/a(1-\gamma_3/\gamma_1))xy\|_{\mathbb{A}_E},$$

and $C$, $D$ are as before. Noting that when $\gamma_3/\gamma_1 \longrightarrow 1$ the quotient $y/(1-\gamma_3/\gamma_1)$ tends to $\ell$, the limit is

$$\tfrac{1}{2}\ell^{-2}\iiint \phi^K\left(n^{-\sigma}a^{-\sigma}\begin{pmatrix}1&&\\&\alpha&\\&&1\end{pmatrix}n^F an\right)\|\tfrac{c}{a}\|^{1/\ell}_{\mathbb{A}_E}\ (\log A\log CD-\tfrac{1}{2}(\log C)^2-\tfrac{1}{2}(\log D)^2),$$

where now

$$A = \|n^F\tfrac{c}{a}\ \ell(1-N\alpha^{-1})\|, \qquad C = \|(1,n_1)(1-N\alpha^{-1})\|, \qquad D = \|(1,n_3)(1-N\alpha^{-1})\|.$$

The second case in the proof of the lemma is when $N\gamma$ approaches a scalar. We may assume that $N\gamma \longrightarrow 1$ since $\phi$ tranforms under $Z(\mathbb{A}_E)$ by a character, and that in fact $\gamma \longrightarrow 1$. When $\gamma$ is sufficiently close to $1$ we may assume that $\gamma$ lies in $A_0(\mathbb{A})$.

We change $n$ so that $an$ is replaced by $na$, write $n^F n$ for $n$ with $n^F$ in $N_0(\mathbb{A})$, $n$ in $N_0(\mathbb{A})\backslash N_0(\mathbb{A}_E)$, and change $na$ back to $an$. The $n_i$ $(i=1,2,3)$ in the weight factor are now replaced by

$$n_1 + n_1^F b/a, \qquad n_2 + n_2^F c/a + n_3 n_1^F b/a, \qquad n_3 + n_3^F c/b,$$

and the integrand has to be multiplied by $\|c/a\|_{\mathbb{A}_E}^{2/\ell}$.

Since $n^F$ and $\gamma$ have coefficients in $\mathbb{A}$ and $\mathbb{A}^\times$ we may change variables as in 2.3.2, replacing $(n^F)^{-1}\gamma n^F$ by $\gamma n^F$. The $n_i^F$ in the weight factor have to be replaced by

$$-n_1^F\gamma_1/\gamma_2 x', \qquad (n_2^F - n_1^F n_3^F/x')/y', \qquad n_3^F/z',$$

where

$$x' = 1 - \gamma_1/\gamma_2, \qquad y' = 1 - \gamma_3/\gamma_1, \qquad z' = 1 - \gamma_3/\gamma_2.$$

We obtain $\frac{1}{4}\ell^{-2}$ times the triple integral of $\phi^K(n^{-\sigma}a^{-\sigma}\gamma n^F an)\|c/a\|_{\mathbb{A}_E}^{2/\ell}$ weighted by (4) of 2.3.2 with

$$A = \|(1, n_3 + n_3^F c/bz', n_2 - n_1 n_3 + n_2^F c/ay' - n_1 n_3^F c/bz' - n_1^F n_3^F c/ay'z')yz\|_{\mathbb{A}_E},$$

$$B = \|(1, n_1 - n_1^F\gamma_1 b/\gamma_2 ax', n_2 - n_3 n_1^F \gamma_1 b/\gamma_2 ax' + n_2^F c/ay' - n_1^F n_3^F c/ax'y')xy\|_{\mathbb{A}_E},$$

$$C = \|(1, n_1 - n_1^F\gamma_1 b/\gamma_2 ax')x\|_{\mathbb{A}_E}, \qquad D = \|(1, n_3 + n_3^F c/bz')z\|_{\mathbb{A}_E}.$$

As $\gamma \longrightarrow 1$ each of $x/x'$, $y/y'$, $z/z'$ tends to $\ell$, and the limit is obtained by replacing $\gamma$ by the limit value 1, and the new A, B, C, D given by

$$A = B = \|\ell^2 n_1^F n_3^F c/a\|_{\mathbb{A}_E}, \qquad C = \|\ell n_1^F b/a\|_{\mathbb{A}_E}, \qquad D = \|\ell n_3^F c/b\|_{\mathbb{A}_E}.$$

The proof of Lemma 6 is now complete. The limits of $J(N\gamma,\phi)$ will be related below to the explicit values of $J_o(\phi)$ for the singular classes $o$.

Finally, note that an alternative definition of the correction is given by replacing $H^E(wg)$ by $H^E(wg) - \tilde{H}^E(w\rho)$ in the original definition of $v_0^E(g)$; cf. end of 2.3.2 and 3.2.1.

### 3.2.3 Singular twisted classes

As in the non-twisted case we state an explicit expression of $J_{\mathcal{O}}(\phi)$ for $\sigma$-equivalence classes $\mathcal{O}$ which contain a diagonal element whose norm is not regular.

LEMMA 7. *The sum of* $J_{\mathcal{O}}(\phi)$ *over the* $\mathcal{O}$ *which contain* $\gamma = \mathrm{diag}(\gamma_1,\gamma_2,\gamma_3)$ *such that* $N\gamma$ *has exactly two equal eigenvalues is equal to three times the sum of*: (1) *the sum over* $\gamma$ *with* $\gamma_1 = \gamma_3$ *in* $A_0^{1-\sigma}(E)Z(E)\backslash A_0(E)$ *of*

$$\tfrac{1}{4}\ell^{-2}\iiint \phi^K(n^{-\sigma}a^{-\sigma}\gamma n^F an)[2\sum \log A_v \sum \log C_vD_v - \sum(\log C_v)^2 - \sum(\log D_v)^2]\|\tfrac{c}{a}\|^{1/\ell}$$

$$+\ \tfrac{1}{2}\ell^{-1}(\lambda_0/\lambda_{-1} - \sum L_v'(1)/L_v(1))\iiint \phi^K(n^{-\sigma}a^{-\sigma}\gamma n^F an)(\sum \log C_vD_v)\|\tfrac{c}{a}\|^{1/\ell}$$

[$n$ *in* $(N_1 \cap N_2)(\mathbb{A})\backslash N_0(\mathbb{A}_E)$, $\quad a = \mathrm{diag}\,(a,b,c)$ *in* $Z(\mathbb{A}_E)A_0(\mathbb{A})\backslash A_0(\mathbb{A}_E)$, $n^F$ *in* $(N_1 \cap N_2)(\mathbb{A}) \cong \mathbb{A}$,

$A_v = |n^F \tfrac{c}{a}\ell(1-N\alpha^{-1})|_v$, $\quad C_v = \|(1,n_1)(1-N\alpha^{-1})\|_v$, $\quad D_v = \|(1,n_3)(1-N\alpha^{-1})\|_v$],

(2) *the sum over* $\gamma$ *with* $\gamma_1 = \gamma_2$ *in* $A_0^{1-\sigma}(E)Z(E)\backslash A_0(E)$ *of the product of*

$$\tfrac{1}{2}\ell^{-1}|M_{1\gamma}^{\sigma}(E)A_1(\mathbb{A})\backslash M_{1\gamma}^{\sigma}(\mathbb{A}_E)|$$

*and*

$$\iiint \phi^K(n^{-\sigma}m^{-\sigma}a^{-\sigma}\gamma amn)\sum \log\|(1,n_2,n_3)(1-N(\gamma_3/\gamma_1))\|_v \; dndmda$$

($n$ *in* $N_1(\mathbb{A}_E)$, $a$ *in* $Z(\mathbb{A}_E)A_1(\mathbb{A})\backslash A_1(\mathbb{A}_E)$, $m$ *in* $A_1(\mathbb{A}_E)M_{1\gamma}^{\sigma}(\mathbb{A}_E)\backslash M_1(\mathbb{A}_E)$).

*All sums are taken over a finite set of places which is independent of* $\phi_{v_0}$ *for a fixed place* $v_0$, *or any larger set; it includes all* $v$ *where* $|1-N\alpha|_v$ *or* $|\ell|_v$ *is not* $1$ ($\alpha = \gamma_2/\gamma_1$ *in* (1) *and* $\gamma_3/\gamma_1$ *in* (2)), *or* $\phi_v \neq \phi_v^0$.

In the proof of Lemma 6 of 3.2.2 it was shown that the first displayed line in (1) here is the value of $J(N\gamma,\phi)$ at $N\gamma$ of (1) in Lemma 7. The proof of Lemma 7 is similar to that of Lemma 3 of 2.3.3, and will be given in 3.3.1 below. We shall next describe $J_{o}(\phi)$ for the remaining singular class.

LEMMA 8. *The distribution* $J_{o}(\phi)$ *where* $o$ *is the class of the identity is the sum of* (1) *the integral over* $n^F$ *in* $N_0(\mathbb{A})$, $n$ *in* $N_0(\mathbb{A})\backslash N_0(A_E)$, $a$ *in* $Z(\mathbb{A}_E)A_0(\mathbb{A})\backslash A_0(\mathbb{A}_E)$, *of the product of* $\phi^K(n^{-\sigma}a^{-\sigma}n^F an)\|c/a\|^{2/\ell}$ $(a = \mathrm{diag}(a,b,c))$ *and the sum of*

$$\ell^{-2}[\sum \log|\ell n_1^F \tfrac{b}{a}|_v \sum \log|\ell n_3^F \tfrac{c}{b}|_v + \tfrac{1}{4}(\sum \log|\ell n_1^F \tfrac{b}{a}|_v)^2 + \tfrac{1}{4}(\sum \log|\ell n_3^F \tfrac{c}{b}|_v)^2],$$

$$\ell^{-1}\tfrac{3}{2}(\lambda_0/\lambda_{-1} - \sum L_v'(1)/L_v(1)) \sum \log|\ell^2 n_1^F n_3^F c/a|_v,$$

$$\frac{3}{2}\sum_v \frac{L_v'(1)}{L_v(1)}\Big(\sum_{w\neq v}\frac{L_w'(1)}{L_w(1)} - \frac{2\lambda_0}{\lambda_{-1}}\Big) + \Big(\frac{\lambda_0}{\lambda_{-1}}\Big)^2 + \frac{\lambda_1}{\lambda_{-1}} + \sum L_v(1)[\tfrac{1}{2}(L_v(s)^{-1})''_{s=1} - (L_v(s)^{-1})'_{s=1}]$$

*and*

$$(2) \qquad 3\ell^{-1}\iiint \phi^K(n^{-\sigma}a^{-\sigma}u(c)an)\|\beta c/\alpha\|^{2/\ell}\log\|\beta c/\alpha\|\, d^\times c\, da\, dn$$

($c$ *in* $\mathbb{A}^\times$, $a = \mathrm{diag}(\alpha,\alpha,\beta)$ *in* $Z(\mathbb{A}_E)A_1(\mathbb{A})\backslash A_1(\mathbb{A}_E)$, $n$ *in* $N_0(\mathbb{A})\backslash N_0(\mathbb{A}_E)$)

*and*

$$(3) \qquad |G(F)Z(\mathbb{A})\backslash G(\mathbb{A})| \int_{Z(\mathbb{A}_E)G(\mathbb{A})\backslash G(\mathbb{A}_E)} \phi(g^{-\sigma}g)\,dg.$$

All sums in (1) are taken over a finite set of places (of E) which includes all $v$ where $|\ell|_v \neq 1$ and $\phi_v \neq \phi_v^0$. As was shown in the proof of Lemma 6 in 3.2.2, the integral weighted by the first row of (1) is the value of $J(N\gamma,\phi)$ at $N\gamma = 1$.

We shall now summarize the discussion of the first sum (over $o$) in (1). It is described by Lemmas 1 and 2, and the following.

LEMMA 9. <u>The sum of</u> $J_o(\phi)$ <u>over the twisted classes</u> $o$ <u>which intersect</u> $A_0(E)$ <u>non-trivially is equal to the sum of</u>

$$\ell^{-2} \sum J(N\gamma,\phi) \qquad (\underline{\text{all}}\ \ \gamma\ \ \underline{\text{in}}\ \ Z(E)A_0{}^{1-\sigma}(E)\backslash A_0(E)),$$

<u>where</u> $J(N\gamma,\phi)$ <u>is the corrected twisted weighted orbital integral of</u> $\phi$ <u>at</u> $\gamma$, <u>defined in</u> 3.2.2, <u>the terms described by the second displayed line of</u> (1) <u>and by</u> (2) <u>of Lemma</u> 7, <u>and by the second and third displayed lines in</u> (1), <u>by</u> (2) <u>and by</u> (3) <u>of Lemma</u> 8.

3.2.4 <u>The term</u> $\sum_o I_o(\phi)$

As was noted in 2.3.4, for the comparison of the two trace formulae we shall have to (1) express the global distribution $J(N\gamma,\phi)$ of Lemma 9 in terms of local distributions, as the comparison will be done by means of the orbital integrals of the components $\phi_v$ of $\phi$, (2) take into account the non-invariant part X of $\sum J_\chi(\phi)$, the contribution to the trace formula (1) from the continous spectrum.

Note that the $\sigma$-invariant terms in $\sum_o J_o(\phi)$ are those of Lemma 1, (2) of Lemma 7, (3) of Lemma 8. The difference between the non-invariant part of $\sum_o J_o(\phi)$ and X is invariant, as all other terms in the trace formula are. To express it in terms of a sum over quadratic and split classes $o$ of products of local (invariant) distributions, we shall need the twisted analogue of 2.3.4. The effect of the twisting here can be traced through [3], and we shall content ourselves with stating the analogous results. Complete proofs will be recorded elsewhere. The result will be the existence of invariant distributions $I(N\gamma,\phi)$ (globally and locally) such that $\ell^{-2}\sum J(N\gamma,\phi)$ of Lemma 9, minus $X$, is $\ell^{-2}\sum I(N\gamma,\phi)$, and their expression in local terms.

If $\gamma$ is an element of $G(\mathbb{A}_E)$ whose norm is quadratic then $I(N\gamma,\phi)$ is the difference between $J(N\gamma,\phi)$ and a smooth compactly supported modulo $Z(\mathbb{A}_E)$ function of $G(\mathbb{A}_E)$, which is the sum over $v$ where $\phi_v$ is not spherical of products over all places, and all components except the one at $v$ are of the form $F(N\gamma,\phi_w)$. The local function $I(N\gamma,\phi_v)$ is the difference of $J(N\gamma,\phi_v)$ and a smooth compactly supported modulo $Z(\mathbb{A}_E)$ function of $G(\mathbb{A}_E)$ which vanishes if $\phi_v$ is spherical. The sum of $I(N\gamma,\phi)$ over the quadratic elements is described by Lemma 2 with $I(N(a\gamma),\phi_v)$ replacing $J(N(a\gamma),\phi_v)$, and Lemmas 3 and 4 are valid.

If $\gamma$ lies in $A_0(\mathbb{A}_E)$ then the distribution $I(N\gamma,\phi)$ is given by

$$I(N\gamma,\phi) = J(N\gamma,\phi) - J_1(N\gamma,\Phi_1(\phi)) - \Phi_0(\phi,N\gamma).$$

Here $\Phi_1(\phi)$ is a smooth compactly supported function on $M_1(\mathbb{A}_E)$ modulo $Z(\mathbb{A}_E)$ which is the sum over $v$ where $\phi_v$ is not spherical of products over all places, where all components except the one at $v$ are of the form $\phi^K_{vN_1}$, where

$$\phi^K_{vN_1}(N\gamma) = |N(\gamma_1\gamma_2)/N\gamma_3^2|_v^{\frac{1}{2}}\iiint\phi_v(k^{-\sigma}n^{-\sigma}a^{-\sigma}\gamma n^F ank)|a/c|_v^{-2/\ell}\,da\,dn\,dn^F\,dk$$

($k \in K\cap M_1\backslash K$, $n \in N_1(F_v)\backslash N_1(E_v)$, $a \in Z(E_v)A_1(F_v)\backslash A_1(E_v)$, $n^F \in N_1(F_v)$, $a = (a,a,c)$, $\gamma_i$ are the eigenvalues of $\gamma$). The distribution $J_1$ on $M_1(\mathbb{A}_E)$ is defined analogously to that of $GL(2)$ in 3.2.1. $\Phi_0(\phi)$ is a smooth compactly supported function on $A_0(\mathbb{A}_E)$ modulo $Z(\mathbb{A}_E)$.

The definition of the local distribution $I(N\gamma,\phi_v)$ is analogous, with an additional condition: If $\phi_v$ is spherical then both $\Phi_1(\phi_v)$ and $\Phi_0(\phi_v)$ are 0. Finally a further distribution $I_1(N\gamma,\phi_v)$ on $M_1(E_v)$ is given by the difference of $J_1(N\gamma,\phi_v)$ and the value at $\gamma$ of some smooth compactly supported function of $A_0(E_v)$ modulo $Z(E_v)$. The precise definition will be given in 5.1.2, where it will actually be needed.

The distributions I were introduced in order to have the following lemma.

LEMMA 10. *For any* $\gamma$ *in* $A_0(\mathbb{A}_E)$ *such that* $N\gamma$ *is regular, the global distribution* $I(N\gamma,\phi)$ *is equal to the sum of*

$$\ell^{-2}\lambda_{-1}^{2} \sum_{v_1\neq v_2} \prod_{i=1,2} I_1(N\gamma,\phi^{K}_{v_i N_1}) \prod_{w\neq v_1,v_2} F(N\gamma,\phi_w)$$

*and*

$$\ell^{-2}\lambda_{-1}^{2} \sum_{v} I(N\gamma,\phi_v) \prod_{w\neq v} F(N\gamma,\phi_w).$$

*For any fixed place* $v_0$ *the sums are taken over any set containing a fixed finite set independent of* $\phi_{v_0}$, *uniformely in* $N\gamma$.

As we have already mentioned a proof of this lemma will not be given here, and will be included in a future discussion of the twisted trace formula, parallel to [3].

The explicit expression for $J_{1\mathfrak{o}}(\phi)$ for the singular class $\mathfrak{o}$ of $GL(2,E)$ recalled from 3.2.1, we finally have:

LEMMA 11. *The sum of* $I_{\mathfrak{o}}(\phi)$ *from* (2) *of* 3.1.1 *over all of the classes* $\mathfrak{o}$ *is equal to the sum of* (1) *the terms of Lemma* 1, (2) *the terms of Lemma* 2 *with* $J(N(a\gamma),\phi_v)$ *replaced by* $I(N(a\gamma),\phi_v)$, (3) *the sum over all* $\gamma$ *in* $Z(E)A^{1-\sigma}(E)\backslash A(E)$ *of the two terms of Lemma* 10, (4) *the terms in the second displayed line of* (1), *and of* (2), *in Lemma* 7, *and those of the second and third displayed lines in* (1), *of* (2) *and of* (3) *in Lemma* 8, *and finally* (5) *the sum over all* $\gamma$ *in* $A_1^{1-\sigma}(E)Z(E)\backslash A_1(E)$ *with singular* $N\gamma$ *of* $\delta(N\gamma)$ (*see* (5) *of Lemma* 2.7) *times*

$$(\lambda_0/\lambda_{-1} - \sum L_v'(1)/L_v(1))\Phi_1(\phi)^{K \cap M_1}_{N_0 \cap M_1}(N\gamma)$$

$$+|M_1(F)A_1(\mathbb{A})\backslash M_1(\mathbb{A})| \int_{A_1(\mathbb{A}_E)M_1(\mathbb{A})\backslash M_1(\mathbb{A}_E)} \Phi_1(\phi)(g^{-\sigma}g)dg.$$

Note that $\Phi_1(\phi)^{K \cap M_1}_{N_0 \cap M_1}$ is the sum over $v$ where $\phi_v$ is not spherical of products over all places $w$, where all factors except the one at $v$ is $\phi^K_{wN_0}$.

3.3.1 Proof of Lemma 7

To claculate $J_{\mathcal{O}}(\phi)$ explicitly where the twisted class $\mathcal{O}$ contains an element $\gamma$ such that $N\gamma$ has exactly two equal eigenvalues and they lie in $NE^\times$, it suffices to assume that $\mathcal{O}$ contains $\gamma = \mathrm{diag}(1,\alpha,1)$ with $\alpha$ in $E^\times$ and $N\alpha \neq 1$. The calculation here will be based on the work done in the non-twisted case.

As in 2.4.1 we express the sum over $P$ and the $\sigma$-equivalence classes of the $\gamma$ in (3) as sums over $\sigma$-conjugacy classes in the intersection of $\mathcal{O}$ with the Levi component $M(E)$. We can use the same list for a set of representatives as in the non-twisted case, only that for later purposes we replace $I$, $I_1$, $I_2$ by

$$I\colon\ u_\alpha(1),\ G; \qquad I_1\colon \begin{pmatrix} 1 & \ell^{-2} & \\ & 1 & \\ & & \alpha \end{pmatrix},\ P_1; \qquad I_2\colon \begin{pmatrix} \alpha & & \\ & 1 & \ell^{-2} \\ & & 1 \end{pmatrix},\ P_2.$$

Here

$$u_\alpha(x) = \begin{pmatrix} 1 & & \ell^{-2} \\ & \alpha & \\ & & 1 \end{pmatrix}.$$

Since

$$G^\sigma_{u_\alpha(1)}(E) = A_3(F)(N_0 \quad M_3)(F) \qquad \text{(cf. 2.4.1)},$$

the contribution from $I, I_1, I_2$ is given by (5) of 2.4.1 with $g^{-\sigma}u_\alpha(1)g$ and $H_0^E$ replacing $g^{-1}u(1)g$ and $H$ there, and the domain of integration is the same, with $G(\mathbb{A}_E)$, $Z(\mathbb{A}_E)$ replacing $G(\mathbb{A})$, $Z(\mathbb{A})$.

The decomposition to be used is

$$g = u_1(n^F)\begin{pmatrix} a & & \\ & b & \\ & & c \end{pmatrix} nk, \qquad \text{where} \quad n = \begin{pmatrix} 1 & n_1 & n_2 \\ & 1 & n_3 \\ & & 1 \end{pmatrix},$$

with

$$n_F \text{ in } \mathbb{A};\ n_1, n_3 \text{ in } \mathbb{A}_E;\ n_2 \text{ in } \mathbb{A}_E \bmod \mathbb{A};\ a,b,c \text{ in } \mathbb{A}_E^\times,$$

and the modular function $\|c/a\|^{1/\ell}$.

Changing $a$ to $c/a$ we obtain the integral over $n$ as in the decomposition, and $a = \operatorname{diag}(c/a,b,c)$ in $Z(\mathbb{A}_E)A_0(\mathbb{A})\backslash A_0(\mathbb{A}_E)$, of the product of $\phi^K(n^{-\sigma}a^{-\sigma}u_\alpha(a^F)an)$, $\|aa^F\|^{1/\ell}$ and an integral over $F^\times\backslash\mathbb{A}^\times$, evaluated as in 2.4.1 to be

$$\frac{3}{2}\ell^{-1}\left(\frac{2}{3}(t_1+t_2) + \log\|(1,n_1)\| + \log\|(1,n_3)\|\right).$$

Here the $\ell^{-1}$ appears since the volume (of b/c; see 2.4.1) is taken with respect to E. The remaining integral over $a^F$ in $F^\times\backslash\mathbb{A}^\times$ will be taken later.

With the above changes the calculation of the contribution from the classes $II, \ldots, II_4$ from 2.4.2 carries over to the current situation. Applying the summation formula as there we obtain

$$\frac{3}{2}\ell^{-2}\iint \underset{s=0}{\text{f.p.}}\left\{\int_{\mathbb{A}^\times}\phi^K(n^{-\sigma}a^{-\sigma}u_\alpha(a^F)an)\|aa^F\|^{\ell^{-1}+s}\, d^\times a^F\right\}$$

$$\log\|(1,n_1)\|\,\|(1,n_3)\|\, da\, dn$$

$$-\frac{3}{4}\ell^{-1}\iiint_{\mathbb{A}}\phi^K(n^{-\sigma}a^{-\sigma}u_\alpha(n^F)an)[(\log\|(1,n_1)\|)^2$$

$$+(\log\|(1,n_3)\|)^2]\|a\|^{1/\ell}\, da\, dn\, dn^F.$$

We shall exploit the function $\theta(s,\phi) = \Pi_v \theta(s,\phi_v)$ to find the "finite part" at $s = 0$, where $\theta(s,\phi_v)$ is given by

$$\frac{|(1-h)(1-h^{-1})|_v}{L(1+\ell s, 1_{F_v})} \iiint_{F_v^\times} \phi_v^K(n^{-\sigma}a^{-\sigma}u_\alpha(a^F)an)\|aa^F\|_v^{\ell^{-1}+s}\, d^\times a^F da dn \qquad (h = N\alpha).$$

Noting that the determinant of the transformation $n_i \longrightarrow n_i - n_i^\sigma \alpha$ of $E_v$ over $F_v$ is $1-h$, and that when $E_v/F_v$ is an unramified field extension there is a set of representatives for the $a_v$ with unit entries, a simple calculation shows that for any $v$ for which $\phi_v = \phi_v^0$ and $h$, $1-h$ are units the local factor $\theta(s,\phi_v)$ is the product of the measures of (1) units of $F_v$, (2) those of $E_v$ (squared), (3) the integers of $E_v$(squared), (4) the image in $E_v$ mod $F_v$ of the integers of $E_v$. The product of these numbers over the $v$ is convergent. Hence the product which defines $\theta(s,\phi)$ is absolutely convergent to an analytic function on $\mathrm{Re}\, s > -\ell^{-1}$, and can be differentiated term by term.

The expression whose f.p. is searched is equal to $\lambda_{-1} L(1+\ell s, 1_F)\theta(s,\phi)$. It has a simple pole at $s = 0$ and the constant term of its Laurent expansion is

$$\lambda_{-1}\lambda_0\, \theta(0,\phi) + \ell^{-1}\lambda_{-1}^2\theta'(0,\phi),$$

that is

$$\frac{3}{2}\ell^{-2}\iiint_{\mathbb{A}} \phi^K(n^{-\sigma}a^{-\sigma}u_\alpha(n^F)an)[\textstyle\sum \log C_v D_v \sum \log A_v$$

$$-\frac{1}{2}(\textstyle\sum \log C_v)^2 - \frac{1}{2}(\sum \log D_v)^2]\|\frac{c}{a}\|^{1/\ell}$$

$$+\frac{3}{2}\ell^{-1}(\lambda_0/\lambda_{-1} - \textstyle\sum L_v'(1)/L_v(1))\iiint \phi^K(n^{-\sigma}a^{-\sigma}u_\alpha(n^F)an)(\sum \log C_v D_v)\|\frac{c}{a}\|^{1/\ell} \quad .$$

Here $A_v = |n^F c/a|_v$, $C_v = \|(1,n_1)\|_v$, $D_v = \|(1,n_3)\|_v$, $n$ ranges over $(N_1 \cap N_2)(\mathbb{A})\backslash N_0(\mathbb{A}_E)$, and $a = \mathrm{diag}(a,b,c)$ over $Z(\mathbb{A}_E)A_0(\mathbb{A})\backslash A_0(\mathbb{A}_E)$. The sums are taken over any set of valuations $v$ of $F$ which include all $v$ such that $\phi_w \neq \phi_w^0$ for any place $w$ of

E above v. Increasing the set of v to include all v with $|\ell|_v \neq 1$ or $|1 - N\alpha^{-1}|_v \neq 1$, we may replace $A_v$, $C_v$, $D_v$ by

$$A_v = |n^F \frac{c}{a} \ell(1-N\alpha^{-1})|_v, \quad C_v = \|(1,n_1(1-N\alpha^{-1})\|_v, \quad D_v = \|(1,n_3)(1-N\alpha^{-1})\|_v.$$

Taking into account the fact that the intersection of $o$ with $Z(E)A_0^{1-\sigma}(E)\backslash A_0(E)$ contains 3 elements, the first line of the explicit expression for $J_o(\phi)$ is the value of the continous function $J(N\gamma,\phi)$ at $\mathrm{diag}(1,\alpha,1)$.

### 3.3.2 Final contribution

The contribution from the classes III, $III_1$, $III_2$ is given by (6) of 2.4.4, with $\gamma = \mathrm{diag}(1,1,\alpha)$ $(N\alpha \neq 1, \alpha$ in $F^\times)$ replacing h, $g^{-\sigma}\gamma g$ instead of $g^{-1}hg$, and the domain of integration is $M_{1\gamma}^{\sigma}(E)Z(\mathbb{A}_E)\backslash G(\mathbb{A}_E)$. Writing g as a'am'mnk with n in $N_1(\mathbb{A}_E)$, a in $Z(\mathbb{A}_E)A_1(\mathbb{A})\backslash A_1(\mathbb{A}_E)$, a' in $Z(\mathbb{A})A_1(F)\backslash A_1(\mathbb{A})$, m in $A_1(\mathbb{A}_E)M_{1\gamma}^{\sigma}(\mathbb{A}_E)\backslash M_1(\mathbb{A}_E)$, m' in $M_{1\gamma}^{\sigma}(E)A_1(\mathbb{A})\backslash M_{1\gamma}^{\sigma}(\mathbb{A}_E)$, we obtain the product of the scalar

$$\frac{3}{2}\ell^{-1}\,|M_{1\gamma}^{\sigma}(E)A_1(\mathbb{A})\backslash M_{1\gamma}^{\sigma}(\mathbb{A}_E)|$$

and

$$\iiint \phi^K(n^{-\sigma}m^{-\sigma}a^{-\sigma}\gamma amn) \sum \log\|(1,n_2,n_3)\|\, dndmda \qquad \left(n = \begin{pmatrix} 1 & 0 & n_2 \\ & 1 & n_3 \\ & & 1 \end{pmatrix}\right).$$

Standard arguments show that for any fixed $v_0$ this integral is non-zero only for $\alpha$ in a finite set independent of $\phi_{v_0}$. The sum over v is taken only over a finite set independent of $\phi_{v_0}$; it includes all v with $\phi_v \neq \phi_v^0$. If we make this set so large as to include all v where $|1-N\alpha|_v \neq 1$, then the weight factor can be replaced by the one in the statement of the lemma.

Lemma 8 can also be proved by similarly adapting the method of proof of Lemma 4 from 2.3.3 to the twisted situation, as above. This can be left as an exercise for the reader.

### 3.4.1 Asymptotic behaviour for GL(2,E)

It remains to be shown that the summation formula can be applied to the various terms in Lemma 11, (3) - (5), which involve weighted orbital integrals. It is clear that the question is local, and for the terms of Lemma 10 (= (3) of Lemma 11) we have to consider only the local distributions $J(\phi_v)$ rather than $I(\phi_v)$.

If $v$ is a place of $F$ which splits in $E$ then $E_v$ is a direct sum of $\ell$-copies of $F_v$. Let $f_{1v}$ be a smooth compactly supported function of $G(F_v)$ modulo $Z(F_v)$. It is left invariant by a subgroup $K'_v$ of $K_v$. Let $f_{iv} (2 \leq i \leq \ell)$ be the function which vanishes outside $Z(F_v)K'_v$, which transforms under $Z(F_v)$ by the usual character, and whose value on $K'_v$ is $1/|K'_v|$, where $|K'_v|$ is the volume of $K'_v$. It suffices to consider only $\phi_v = (f_{1v}, f_{2v}, \ldots, f_{\ell v})$. This restriction will simplify the arguments, but will not pose any restriction on the generality of the final results. The proof of Lemma 1.15 shows that $J_1(N\gamma,\phi_v) = \ell^q J_1(h,f_v)$ where $h = N\gamma$ is in $G(F_v)$ and $f_v = f_{1v} * f_{2v} * \ldots * f_{\ell v}$, thus reducing the question of the behaviour of $J_1(N\gamma,\phi_v)$ to that of $J_1(h,f_v)$, which was discussed in 2.7.1 and 2.7.2. Note that the above Lemma 15 applies also to the corrected weighted integrals since if $u$ lies in $F_v^\times$ then $\log|u|_{E_v} = \ell \log|u|_{F_v}$, and the corrections of $J_1(N\gamma,\phi_v)$ and $J_1(h,f_v)$ were done by means of the same element of $F^\times$.

When $E_v$ is a field we have to distinguish between the archimedean and p-adic cases. In the first case $F_v = \mathbb{R}$, $E_v = \mathbb{C}$, $\ell = 2$, and the integrability of the Fourier transform of $J(N\gamma,\phi_v)$ can be dealt with as in the non-twisted case (cf. [12], last line of §9). In the p-adic case we shall have to discuss the asymptotic behaviour of $J(N\gamma,\phi_v)$ near the singular set and make use of Lemma 8 of 2.6.1.

The asymptotic behaviour of $J_1(N\gamma,\phi_v)$ as $N(\gamma_2/\gamma_1) \longrightarrow 1$ was discussed in great detail in [12], §9. The correction and Lemma 1.8 show that it suffices to give a briefer discussion. This is an exercise which is nevertheless quite elaborate even in the case of $GL(2,E)$, and therefore will be given here. It will serve to illustrate the ideas required in the case of $GL(3,E)$.

Since we deal with a local question we may drop $v$, and start with the local analogue of the last displayed expression for $J_1(N\gamma,\phi)$ in 3.2.1. Since $N(\gamma_2/\gamma_1) \longrightarrow 1$ we may assume that $\gamma_2/\gamma_1 \longrightarrow 1$ and that $\gamma_2/\gamma_1$ lies in $F^\times$. Hence $u \longrightarrow 0$ where $u = 1 - \gamma_2/\gamma_1$ lies in $F^\times$, and $|1 - N(\gamma_2/\gamma_1)|$ is equal to $|\ell u|$, and $\Delta(N\gamma) = |\ell u|$. The change $(n^F)^{-1}\gamma n^F \longrightarrow \gamma n^F$ of 3.2.1 has the effect of dividing $\Delta(N\gamma)$ by $|u|$. Then $J_1(N\gamma,\phi)$ is equal to

$$|\ell| \iiint \phi^K(n^{-\sigma}a^{-\sigma}\gamma n^F an) \log\|(\ell u, \ell un+\ell n^F/a)\| \; |a|^{-1},$$

where

$$a = \begin{pmatrix} a' & 0 \\ 0 & a'' \end{pmatrix} \text{ in } Z(E)A(F)\backslash A(E), \; a = a'/a'' \text{ in } F^\times\backslash E^\times, \; n = \begin{pmatrix} 1 & n \\ 0 & 1 \end{pmatrix}, \; n^F = \begin{pmatrix} 1 & n^F \\ 0 & 1 \end{pmatrix},$$

and $n^F$ lies in $F$, $n$ lies in $E$ modulo $F$. As in [12] we denote the valuation with respect to $E$ (resp. $F$) by a double (resp. single) bar.

The domain of integration is divided into two subdomains, where $\|n^F/a\| > \|u(1,n)\|$, and where $\|n^F/a\| \leq \|u(1,n)\|$. We obtain the sum of

$$|\ell| \iiint \phi^K(n^{-\sigma}a^{-\sigma}\gamma n^F an) \log\|\ell n^F/a\| \; |a|^{-1},$$

where we may set $\gamma = 1$ if $\gamma$ is sufficiently close to $1$, and the integral over $\|n^F/a\| \leq \|u(1,n)\|$ of the product of

$$|\ell| \; |a|^{-1}\phi^K(n^{-\sigma}a^{-\sigma}\gamma n^F an)$$

and

$$\log\|(\ell u, \ell un+\ell n^F/a)\| - \log\|\ell n^F/a\|.$$

To continue we have to assume that $a$ and $n$ are taken over a fixed set of representatives modulo $F^\times$ and $F$. If $|u|$ is small $n^F$ can be erased from $\phi^K$, and then $n^F$ can be multiplied by $u$. We obtain the product by $|u|$ of an integral over the domain $\|n^F/a\| \leq \|(1,n)\|$ (condition which is independent of $u$) of $|\ell/a|\phi^K(n^{-\sigma}a^{-\sigma}\gamma an)$ (independent of $u$) multiplied by a weight factor (polynomial in $\log|u|$; in fact a constant in $\log|u|$), as required for the application of Lemma 1.8.

### 3.4.2 Asymptotic behaviour for GL(3,E)

We shall not deal with all cases where the asymptotic behaviour has to be considered, but content ourselves with the discussion of the analogue of the case considered in 2.7.1. It is clear that the methods can be used in the case of the analogue of 2.7.2 and all other cases of interest.

In the situation described by the first case of the proof of Lemma 6, we have the integral over $a$, $n$, $n^F$ of the product of $\phi^K = \phi^K(n^{-\sigma}a^{-\sigma}\gamma n^F an)\|\frac{c}{a}\|^{1/\ell}$ by a weight factor. As in 2.7.1 we take this weight factor to be $(\log A)^2$, where

$$A = \|(1,n_3,n_2-n_1n_3+n^F\tfrac{c}{a}u^{-1})yz\| \qquad (u=1-\gamma_3/\gamma_1,\ y=1-N(\gamma_3/\gamma_1)).$$

As $\gamma = (\gamma_1,\gamma_2,\gamma_3) \longrightarrow (1,\alpha,1)$, $z = 1 - N(\gamma_3/\gamma_1) \longrightarrow 1-\alpha^{-\ell}$ and hence $z$ can be considered to be the constant $1-\alpha^{-\ell}$ in $A$ when $\gamma$ is close enough to the limit. Further, $|y| = |\ell u|$ for such $\gamma$. Since we may write $\phi^K = \phi_1^K + \phi_2^K$ with $\phi_1^K$ supported on $|n_3| \leq 1$ and $\phi_2^K$ supported on $1 < |n_3|$, we may assume that $\phi^K = \phi_1^K$ and omit the similar case where $\phi^K = \phi_2^K$.

Splitting the domain of integration into two, depending on whether $\|n^F \frac{c}{a}\|$ is $\leq$ or $>$ then $\|u(1,n_2-n_1n_3)\|$, the integral becomes the sum of the smooth

$$\iiint \phi^K \|\frac{c}{a}\|^{1/\ell} (\log\|\ell z n^F c/a\|)^2 da\,dn\,dn^F$$

and the integral over $\|n^F c/a\| \leq \|u(1,n_2-n_1n_3)\|$ of $\phi^K\|c/a\|^{1/\ell}$ weighted by

$$(\log\|\ell z(u,u(n_2-n_1n_3)+n^F c/a)\|)^2 - (\log\|\ell z n^F c/a\|)^2.$$

If $|u|$ is small $\phi^K$ is independent of $n^F$ (which is erased), then $n^F$ in $F$ is multiplied by $u$ and the last integral is clearly the product of $|u|$ and a linear function in $\log u$, as required.

## §4. THE CONTINUOUS SPECTRUM

### 4.1.1 Notations

The difference between $\ell$ times the twisted trace formula ((2), chapter 3), and the trace formula ((2), chapter 2) can be written in the form

$$\ell \operatorname{tr} r(\phi') - \operatorname{tr} r(f) + \ell \sum I_\chi(\phi) - \sum I_\chi(f) = \ell \sum I_o(\phi) - \sum I_o(f).$$

In the last two chapters some of the properties of the sums on the right were studied. In the next chapter we shall prove that for $\phi = \otimes\phi_v$ and $f = \otimes f_v$ with $\phi_v \longrightarrow f_v$ for all $v$ both sides are 0. In the present chapter we shall describe $\sum I_\chi(f)$ and $\ell \sum I_\chi(\phi)$ and their difference. This will be of use not only in the next chapter but also in the final one, where our applications will be deduced from the resulting equality of traces. Our discussion is based on Arthur [2,3].

We denote by P, M, A, N, K a parabolic subgroup of G, the Levi component of P, the split component of M, the unipotent radical of P and a special maximal compact subgroup (respectively). Given a unitary representation $\pi$ of $M(F)\backslash M(\mathbb{A})$, which transforms under $A(\mathbb{A})$ by a character, on a Hilbert space H , and we let $H_P(\pi)$ be the Hilbert space of measurable functions $\psi: G \longrightarrow H$ with

$$\psi(mng) = \delta_P(m)^{1/2}\pi(m)\psi(g) \quad (n \text{ in } N(\mathbb{A}),\ m \text{ in } M(\mathbb{A}),\ g \text{ in } G(\mathbb{A}))$$

and

$$\|\psi_c\|^2 = \int_K \int_{A(\mathbb{A})M(F)\backslash M(\mathbb{A})} |(\psi(mk),c)|^2 dm dk < \infty \quad ;$$

here $\delta_P$ denotes the modular function on $P$, $c$ is any vector in $H$ and the inner product in $H$ is denoted by brackets. The induced representation $I_P(\pi)$ of $G$ on $H_P(\pi)$ is given by

$$(I_P(\pi,h)\psi)(g) = \psi(gh) \quad (\psi \text{ in } H_P(\pi);\ g,h \text{ in } G(\mathbb{A})).$$

The space of functions on $A(\mathbb{A})M(F)\backslash M(\mathbb{A}) \times K$ obtained by restricting functions in $H_P(\pi)$ is also denoted by $H_P(\pi)$, and the induced representation on this new space is again $I_P(\pi)$; it is unitary since $\pi$ is unitary. For any vector $\Lambda$ in the real vector space $X(A(\mathbb{A})M(F)\backslash M(\mathbb{A}))_F \otimes \mathbb{R}$ we put

$$\pi_\Lambda(m) = \pi(m)\exp(\Lambda(H_M(m)))$$

where $H_M$ is the homomorphism from $A(\mathbb{A})M(F)\backslash M(\mathbb{A})$ to $A_P = \mathrm{Hom}(X(M)_F,\mathbb{R})$ of Chapter 2. We note that $H_P(\pi_\Lambda) = H_P(\pi)$.

For any parabolic subgroups $P$, $P'$ of $G$ such that $A, A'$ contain $A_0$ we let $W(A,A')$ denote the set of isomorphisms from $A = A_P$ onto $A' = A_{P'}$ obtained by restricting elements of the Weyl group $W$ of $A_0$ in $G$ to $A$. The $P$ and $P'$ are said to be associated if $W(A,A')$ is not empty. For each $s$ in $W$ we fix a representative $w_s$ in the normalizer of $A_0$ in $K \cap G(F)$. The pairs $(M,\pi)$ and $(M',\pi')$ of representations $\pi,\pi'$ of Levi subgroups $M,M'$ (resp.) are said to be equivalent if there is an $s$ in $W(A,A')$ such that the representation

$$(s\pi)(m') = \pi(w_s^{-1}m'w_s) \qquad (m' \text{ in } M'(\mathbb{A}))$$

of $M$ is unitarily equivalent to $\pi'$. The set of equivalence classes of pairs is denoted by $X$ and for each $\chi$ in $X$ there is a class $P_\chi$ of associated parabolic subgroups.

For associated $P$ and $P'$ the integral

$$(M_{P'|P}(\pi_\Lambda)\psi)(g) = \int_{N\cap N'(\mathbb{A})\backslash N'(\mathbb{A})} \psi(ng)dn$$

converges for $\Lambda$ in the positive Weyl chamber of $A_P$ and defines an intertwining operator from $H_P(\pi_\Lambda)$ to $H_{P'}(\pi'_\Lambda)$ with $\pi'_\Lambda = s\pi_\Lambda$. The Eisenstein series was defined by

$$E(g,\psi,\pi_\Lambda) = \sum_{\delta \text{ in } P(F)\backslash G(F)} \psi(\delta g) \quad (\psi \text{ in } H_P(\pi_\Lambda)).$$

For each $\chi$ in $X$ we let $H_P(\pi)_\chi$ be $H_P(\pi)$ if $(M,\pi)$ is in $\chi$, and the empty set otherwise. Let $B_P(\pi)_\chi$ be an orthonormal basis for $H_P(\pi)_\chi$. We shall denote by $X_c$ the set of all $\chi$ in $X$ such that if $(P,\pi)$ lies in $\chi$ then either $P \neq G$ or $\pi$ is one-dimensional.

4.1.2 Kernels

Langland's theory of Eisenstein series shows that a kernel for the restriction of $r(f)$ to the continuous spectrum (non-cuspidal part of $L^2(\omega)$) is given by

$$\sum_{\chi\in X_c} \sum_{P\in P_\chi} n(A)^{-1}(2\pi)^{-|A/Z|} \int_{A_P} \sum_\pi \{\sum_{\alpha,\beta} (I_P(\pi_{i\Lambda},f)\psi_\beta,\psi_\alpha)$$

$$E(g,\psi_\alpha,\pi_{i\Lambda})\overline{E}(h,\psi_\beta,\pi_{i\Lambda})\}\, d\Lambda .$$

Here $n(A)$ denotes the number of Weyl chambers (connected components) in the complement of $\mathcal{A}_P$ in $\mathcal{A}_{P_0}$; the sum over $\pi$ is taken over a set of representatives for the orbits $\{\pi_{i\Lambda};\ \Lambda \text{ in } \mathcal{A}_P\}$; the inner sum is over all $\psi_\alpha$ and $\psi_\beta$ in $B_P(\pi)_\chi$. The restriction of the function can be modified and then integrated over $Z_E(\mathbb{A})G(F)\backslash G(\mathbb{A})$. The result (see [1]) is $\sum_{\chi\in X_c} J_\chi(f)$.

Using Langland's inner product formula Arthur [2] obtained a useful expression for each of the $J_\chi(f)$. To record it here we need more preparations. Any group $A$ which is conjugate in $G$ to one of $Z$, $A_1$, $A_0$ is called special. For a pair $A,A'$ of special subgroups with $A \supset A'$ let $W^{M'}(\mathcal{A})$ be the subset of $W(\mathcal{A},\mathcal{A})$ whose space of fixed vectors in $\mathcal{A}$ is $\mathcal{A}'$. If $P'$ is a parabolic subgroup with a split component $A'$ and $Q$ is a parabolic subgroup of $M'$ with split component $A$ then there is a unique parabolic subgroup $P'(Q)$ $(= QN')$ of $G$ with split component $A$ which is contained in $P'$ and its intersection with $M'$ is $Q$. Then for any $\Lambda$ in $\mathcal{A}'$ the limit

$$M(P,A',\pi_{i\Lambda}) = \lim_{\lambda\to 0} \sum_{P'} M_{P'(Q)|P}(\pi_{i\Lambda})^{-1} M_{P'(Q)|P}(\pi_{i(\Lambda+\lambda)})/\Pi_{\alpha \text{ in } \Delta_{P'}} \langle\lambda,\alpha\rangle \qquad (1)$$

exists as an operator on $H_P(\pi_{i\Lambda}) = H_P(\pi)$.

For any $\chi$ in $X_c$ the formula for $J_\chi(f)$ is given by the sum over all special subgroups $A'$ and $A$ of $G$ with $A' \subset A \subset A_0$ and over all $s$ in $W^{M'}(A)$ of

$$c_s \int_{A'} \sum_\pi \{\sum_{\alpha,\beta} (I_P(\pi_{i\Lambda}, f)\psi_\beta, \psi_\alpha)(M(P,A',\pi_{i\Lambda})M(s,\pi)\psi_\alpha, \psi_\beta)\} d\Lambda \qquad (2)$$

where

$$c_s^{-1} = n^{M'}(A_0) n(A')(2\pi)^{-|A'/Z|} |\det(1-\mathrm{ad}(s))_{A/A'}| .$$

Here $n^{M'}(A_0)$ is the number of Weyl chambers in the complement of $\mathfrak{A}_{P'}$ in $\mathfrak{A}_{P_0}$, and the operator $M(s,\pi)$ on $H_P(\pi)$ is defined by

$$(M(s,\pi)\psi)(g) = \int_{N(\mathbb{A}) \cap w_s N(\mathbb{A}) w_s^{-1} \backslash N(\mathbb{A})} \psi(w_s^{-1} ng)\, dn \quad (\psi \text{ in } H_P(\pi)).$$

Jacquet has proved (private communication) that there are meromorphic functions $m_{P'|P}(\pi_\Lambda)$ of $\Lambda$ in $\mathfrak{A}_o \otimes \mathbb{C}$ such that the normalized intertwining operators

$$R_{P'|P}(\pi_\Lambda) = m_{P'|P}(\pi_\Lambda)^{-1} M_{P'|P}(\pi_\Lambda)$$

have the properties assumed by Arthur in [3]. In [3] the invariant distributions $I_\chi(f)$ were defined by (2) where $M(P,A',\pi_{i\Lambda})$ is replaced by the scalar $m(P,A',\pi_{i\Lambda})$. This scalar is the logarithmic derivative defined by (1) with $M_{P'(Q)|P}(\pi_{i\Lambda})$ replaced by $m_{P'(Q)|P}(\pi_{i\Lambda})$. The invariant distributions $I_\chi(f)$ will be described in more detail below and then compared with the contributions $I_\chi(\phi)$ from the continuous spectrum of the twisted trace formula.

A kernel for the restriction of $r(\phi')$ to the continuous spectrum (non-cuspidal part) of $L^2(\omega_E)$ is given by

$$\sum_{\chi\in X_c}\sum_{P\in P_\chi} n(A)^{-1}(2\pi)^{-|A/Z|}\int_{A_P}\sum_{\pi^E}\{\sum_{\alpha,\beta}(I_P(\pi^E_{i\Lambda},\phi)\psi_\beta,\psi_\alpha)$$

$$E(g,{}^\sigma\psi_\alpha,{}^\sigma\pi^E_{i\Lambda})\,\overline{E}(h,\psi_\beta,\pi^E_{i\Lambda})\}d\Lambda\ ,$$

where all symbols are the same as in the previous case but they are defined with respect to E. We denote by ${}^\sigma\psi$ (and ${}^\sigma\pi^E$) the function (and representation) defined by ${}^\sigma\psi(g) = \psi(g^\sigma)$ and ${}^\sigma\pi^E(g) = \pi^E(g^\sigma)$).

Using Arthur's arguments and Langland's inner product formula one shows as before that for each $\chi$ in $X_c$ the contribution $J_\chi(\phi)$ is given by the sum over $A',A$ with $A'\subset A\subset A_0$ and $s$ in $W^{M_1}(A)$ of

$$c_s\int_{A'}\sum_{\pi^E}\{\sum_{\alpha,\beta}(I_P(\pi^E_{i\Lambda},\phi)\psi_\beta,\psi_\alpha)(M_E(P,A',{}^\sigma\pi^E_{i\Lambda})M(s,{}^\sigma\pi^E)I_P(\sigma,\pi^E)\psi_\alpha,\psi_\beta)\}d\Lambda$$

$$= c_s\int_{A'}\sum_{\pi^E}\{\sum_\beta(M_E(P,A',{}^\sigma\pi^E_{i\Lambda})M(s,{}^\sigma\pi^E)I_P(\sigma,\pi^E)I_P(\pi^E_{i\Lambda},\phi)\psi_\beta,\psi_\beta)\}d\Lambda \qquad (3)$$

where the inner sum is taken over all $\psi_\alpha, \psi_\beta$ (or $\psi_\beta$) in $B_P(\pi^E)_\chi$. Here we put $I_P(\sigma,\pi^E)\psi_\alpha = {}^\sigma\psi_\alpha$. The operator $M_E(P,A',{}^\sigma\pi^E_{i\Lambda})$ is defined by the formula (1) with respect to E. The normalizing factor $m_{P'|P}(\pi^E_\Lambda)_E$ can be introduced in the present case.

The invariant distributions $I_\chi(\phi)$ are defined by (3) where we replace the operator $M_E(P,A',{}^\sigma\pi^E_{i\Lambda})$ by the scalar valued function $m_E(P,A',{}^\sigma\pi^E_{i\Lambda})$. This in turn is defined by (1) with $M_{P'(Q)|P}(\pi_{i\Lambda})$ being replaced by $m_{P'(Q)|P}(\pi^E_{i\Lambda})_E$.

4.2.1 The $I_\chi(f)$ and $I_\chi(\phi)$

We shall now describe in more detail each of the contributions $I_\chi(f)$ and $I_\chi(\phi)$, arranging them according to the classes $\chi$, and then calculate the difference $\ell\sum I_\chi(\phi) - \sum I_\chi(f)$ if $\phi = \otimes\phi_v$, $f = \otimes f_v$ and $\phi_v \longrightarrow f_v$ for all $v$. We note again that $B_P(\pi)_\chi$ is empty unless P lies in the associated class $\mathcal{P}_\chi$ of $\chi$. We shall then write our list in terms of $\mathcal{P}_\chi = \{P\}$. In our case of $G = GL(3)$ there are three possibilities for the classes They will be given in the cases (a), (b), (c) below.

4.2.1 (a) $\mathcal{P}_\chi = \{G\}$

Since now $A = Z$ we must have $A' = Z$ and $P' = G$. The set $W^{M'}(A)$ contains only the identity, $c_s = 1$, and $A'$ is a single point. The term $I_\chi(f)$ becomes the sum of $\operatorname{tr}\pi(f)$ over all one-dimensional representations $\pi$ of $G(\mathbb{A})$ in $L^2(\omega)$. Similarly the term $I_\chi(\phi)$ is equal to

$$\sum_{\pi^E} \sum_\beta (\pi^E(\sigma)\pi^E(\phi)\psi_\beta,\psi_\beta) = \sum_{\pi^E} \mathrm{tr}\{\pi^E(\sigma)\pi^E(\phi)\} ,$$

where the $\pi^E$ are the one-dimensional representations of $G(\mathbb{A}_E)$ in $L^2(\omega_E)$.

The non-zero terms on the right are only those for which $^\sigma\pi^E = \pi^E$. Now if $\pi^E(g) = \mu^E(\deg g)$ where $\mu^E$ is a character of $\mathbb{A}_E^\times$ with $^\sigma\mu^E = \mu^E$ then there exists a character $\mu$ of $\mathbb{A}^\times$ with $\mu^E(x) = \mu(Nx)$. Throughout this chapter we denote by $\zeta$ a non-trivial character $F^\times N\mathbb{A}_E^\times\backslash\mathbb{A}^\times$. Each of the representations $\pi(g) = \zeta^j\mu(\det g)$ $(0 \leq j < \ell)$ contributes a term to $I_\chi(f)$; for each of them we have that $\mathrm{tr}\,\pi(f)$ is equal to $\mathrm{tr}\,\pi^E(\phi')$. Since there are $\ell$ such $\pi$'s which map to $\pi^E(g) = \mu^E(\det g)$ we conclude that

$$I_\chi(f) = \ell I_\chi(\phi)$$

for all classes $\chi$ with $P_\chi = \{G\}$ .

4.2.1 (b) $P_\chi = \{P_1\}$

Since for $P = P_1$ the $W^G(A)$ is empty the first sums in (2) and (3) reduce to the single term with $A' = A$. The $W^M(A)$ now contains only the identity. The value of the constant $c_s$ is $1/8\pi$, and since $\psi = \sum(\psi,\psi_\alpha)\psi_\alpha$ we obtain (from (2))

$$(8\pi)^{-1}\int_{A'} \sum_\pi m(P,A',\pi_{i\Lambda})\mathrm{tr}\{I_P(\pi_{i\Lambda},f)\}d\Lambda \qquad (4)$$

Similarly we find that $I_\chi(\phi)$ is equal to

$$(8\pi)^{-1}\int_{A'} \sum_{\pi^E} m_E(P,A',{}^\sigma\pi^E_{i\Lambda})\mathrm{tr}\{I_P(\sigma,\pi^E)I_P(\pi^E_{i\Lambda},\phi)\}d\Lambda .$$

For the $\pi^E$ which appear here non-trivially we have ${}^{\sigma}\pi^E = \pi^E$. Each $\pi$ is a pair $(\rho,\eta)$ where $\rho$ is either a cuspidal or a one-dimensional representation of $GL(2,\mathbb{A})$, and $\eta$ is a character of $F^\times\backslash\mathbb{A}^\times$.

The theory of base change for $GL(2)$ [12] establishes that for any $\pi^E$ with ${}^{\sigma}\pi^E = \pi^E$ there exists a $\pi = (\rho,\eta)$ in (4) which corresponds to $\pi^E$, where $\rho$ is not of the form $\pi(\tau)$ with $\tau = \mathrm{Ind}(W_{E/F}, W_{E/E}, \eta_1)$ for any character $\eta_1$ of $\mathbb{A}_E^\times$. More precisely we have

$$\mathrm{tr}\{I_P(\sigma,\pi^E)I_P(\pi^E_{i\Lambda},\phi)\} = \mathrm{tr}\ I_P(\pi_{i\Lambda},f) \quad .$$

Each of the representations $\pi_{ij} = (\zeta^i\rho,\zeta^j\eta)$ $(0 \leq i,j < \ell)$ contributes a term of $I_\chi(f)$ and it corresponds to $\pi^E$. These are all of the $\pi$ which correspond to $\pi^E$. The identity $\Pi m_{P'|P}(\pi_\Lambda) = m_{P'|P}(\pi^E_\Lambda)_E$ (product over all $\pi$ which correspond to $\pi^E$), which follows from the analogous property of the L-functions defining these m-factors, and the fact that $m_E(P,A',\pi^E_{i\Lambda})$ is a logarithmic derivative which is defined with respect to the extension $E$ of degree $\ell$, imply that

$$\sum m(P,A',\pi_{i\Lambda}) = \ell m_E(P,A',\pi^E_{i\Lambda}), \tag{5}$$

where the sum is taken over all $\pi$ which correspond to $\pi^E$.

In this way we obtained each $\pi = (\rho,\eta)$ in (4), with $\rho$ not of the form $\pi(\tau)$, from the $\pi^E$ of $I_\chi(\phi)$. We deduce that $\ell I_\chi(\phi)$ is equal to the subsum of (4) taken over all $\pi = (\rho,\eta)$ with $\rho \neq \pi(\tau)$ for all $\tau$. The remaining terms from $I_\chi(f)$ will be considered below.

4.2.1 (c) $\mathcal{P}_\chi = \{P_0\}$ , $A' = A_0$

Here $A = A_0$ and the first sum in (3) has three terms. We shall treat each of them separately. We begin with the term corresponding to $A' = A_0$. In this case $W^{M'}(A)$ consists of a single element, the identity. We write $c$ for $c_{id}$. The corresponding contribution to $I_\chi(f)$ is

$$c\int_{A'} \sum_\pi m(P,A',\pi_{i\Lambda})\mathrm{tr}\{I_P(\pi_{i\Lambda},f)\} . \tag{6}$$

Similarly we see that the contribution to $I_\chi(\phi)$ corresponding to $A' = A_0$ is

$$c \int_{A'} \sum_{\pi^E} m_E(P,A',{}^\sigma\pi^E_{i\Lambda})\mathrm{tr}\{I_P(\sigma,\pi^E)I_P(\pi^E_{i\Lambda},\phi)\}d\Lambda.$$

The $\pi^E$ which contribute a non-zero term to the sum here are those with ${}^\sigma\pi^E = \pi^E$. These $\pi^E$ are triples $(\eta_1^E,\eta_2^E,\eta_3^E)$ of characters of $E^\times\backslash\mathbb{A}_E^\times$ with ${}^\sigma\eta_i^E = \eta_i^E$ $(1 \leq i \leq 3)$. For each such $\eta_i^E$ there exists a character $\eta_i$ of $\mathbb{A}^\times$. Each $\pi = (\zeta^{i_1}\eta_1,\zeta^{i_2}\eta_2,\zeta^{i_3}\eta_3)$ $(0 \leq i_j < \ell,\ 1 \leq j \leq 3)$ appears in (6) and it corresponds to $\pi^E$, that is

$$\mathrm{tr}\, I_P(\pi_{i\Lambda},f) = \mathrm{tr}\{I_P(\sigma,\pi^E)I_P(\pi^E_{i\Lambda},\phi)\} .$$

These are all of the $\pi$ which correspond to $\pi^E$. Since now the $\ell$-th power of $m_{P'|P}(\pi_\Lambda)$ is equal to $m_{P'|P}(\pi^E_\Lambda)_{E'}$ as $m_E(P,A',\pi^E_{i\Lambda})$ is a second derivative and it is defined with respect to $E$ we have that (5) is valid in this case. The conclusion is that the contribution to $I_\chi(f)$ corresponding to $A' = A_0$ is equal to $\ell$ times the contribution to $I_\chi(\phi)$ corresponding to $A' = A_0$.

4.2.1 (c) $p_\chi = \{P_0\}$, $A' = A_1$

We can now continue with the term corresponding to $A' = A_1$. In this case $W^{M'}(A_0)$ contains the unique element $s = s_{\alpha_1}$; we have $c_s = 1/16\ \pi$. The contribution to $I_\chi(f)$ from such $A'$ is

$$(16\ \pi)^{-1}\int_{A'} \sum_\pi m(P_0,A,\pi_{i\Lambda})\mathrm{tr}\{M(s_{\alpha_1},\pi)I_P(\pi_{i\Lambda},f)\}d\Lambda\ . \tag{7}$$

The $\pi$ which contribute a non-zero term to this sum satisfy $\eta_1 = \eta_2$ if $\pi$ is the triple $(\eta_1,\eta_2,\eta_3)$ of characters on $\mathbb{A}^\times$ (<u>not</u> $F^\times N\mathbb{A}_E^\times$). Since $M(s_{\alpha_1},\pi)$ intertwines the irreducible $I_P(\pi_{i\Lambda})$ with itself it is a scalar which is denoted by $m(s_{\alpha_1},\pi)$ and written outside the trace symbol.

The corresponding contribution to $I_\chi(\phi)$ is given by

$$(16\ \pi)^{-1}\int_{A'} \sum_{\pi^E} m_E(P_0,A',{}^\sigma\pi^E_{i\Lambda})\mathrm{tr}\{M(s_{\alpha_1},{}^\sigma\pi^E)I_P(\sigma,\pi^E)I_P(\pi^E_{i\Lambda},\phi)\}d\Lambda.$$

Here the sum contains only $\pi^E$ with ${}^\sigma\pi^E = s_{\alpha_1}\pi^E$, namely the $\pi^E$ of the form $(\eta_1^E,\eta_2^E,\eta_3^E)$ with ${}^\sigma\eta_1^E = \eta_2^E$, ${}^\sigma\eta_2^E = \eta_1^E$ and ${}^\sigma\eta_3^E = \eta_3^E$.

Consider first the $\pi^E$ with $\eta_1^E = \eta_2^E$. In this case $M(s_{\alpha_1},\pi^E)$ $(= M(s_{\alpha_1},{}^\sigma\pi^E))$ is a scalar since it intertwines the irreducible $I_P(\pi^E_{i\Lambda})$ with itself. This scalar, which we may denote by $m(s_{\alpha_1},\pi^E)$, will be written outside the trace symbol. But for $\pi = \pi(\eta_1,\eta_2,\eta_3)$ the $m(s_{\alpha_1},\pi)$ is given by the quotient of $L(1,\eta_2/\eta_1)$ by $L(1,\eta_1/\eta_2)$ if $\eta_1 \neq \eta_2$. The value at $\eta_1 = \eta_2$ has to be evaluated as a limit, hence it is

$$\lim_{t\to 0} m(s_{\alpha_1}, \pi(\eta_1\alpha^t, \eta_2\alpha^{-t}, \eta_3)) = \lim_{t\to 0} L(1-2t, 1_F)/L(1+2t, 1_F) = -1$$

(where $\alpha(x) = |x|$). Since $m(s_{\alpha_1}, \pi^E)$ for $\pi^E = (\eta_1^E, \eta_2^E, \eta_3^E)$ is the same object defined with respect to a different field we have that its value is also $-1$ whenever $\eta_1^E = \eta_2^E$.

For each $\pi^E$ under consideration there is a pair $\eta_1$, $\eta_3$ of characters of $F^\times\backslash\mathbb{A}^\times$ with $\eta_i^E(x) = \eta_i(Nx)$ $(i = 1,3)$. Each $\pi = (\zeta^i\eta_1, \zeta^i\eta_1, \zeta^j\eta_3)$ $(0 \leq i,j < \ell)$ contributes a term to (7) and it corresponds to $\pi^E$, that is

$$\operatorname{tr} I_P(\pi_{i\Lambda}, f) \quad = \operatorname{tr}\{I_P(\sigma, \pi^E) I_P(\pi_{i\Lambda}^E, \phi)\}.$$

Each $\pi$ which corresponds to $\pi^E$ is of this form. There are $\ell^2$ such $\pi$ which correspond to $\pi^E$, and the logarithmic derivatives $m$ and $m^E$ satisfy the identity (5). Every $\pi$ in (7) is obtained from $\pi^E$ in the above way. We conclude that the contribution to $I_\chi(f)$ from $A'$ with $A_0 \neq A' \neq Z$ is equal to $\ell$ times the contribution to $I_\chi(\phi)$ from such $A'$ and the $\pi^E$ with $\eta_1^E = \eta_2^E$.

We have to consider also the terms contributed to $I_\chi(\phi)$ from the remaining $\pi^E$'s for which ${}^\sigma\eta_1^E = \eta_2^E$, ${}^\sigma\eta_2^E = \eta_1^E$ and ${}^\sigma\eta_3^E = \eta_3^E$, but such that $\eta_1^E \neq \eta_2^E$. Such $\pi^E$ exists only when $\ell = 2$. As usual we note that there is a character $\eta_3$ of $\mathbb{A}^\times$ with $\eta_3^E(x) = \eta_3(Nx)$. The theory of base change for GL(2) asserts that there is a unique representation of $GL(2,\mathbb{A})$ which corresponds to the representation $GL(2,\mathbb{A}_E)$ induced from the pair

$({}^{}\eta_1^E, {}^\sigma\eta_1^E)$. This is a cuspidal representation of the form $\pi(\tau)$ with $\tau = \mathrm{Ind}(W_{E/F}, W_{E/E}, \eta_1)$, and we have

$$\mathrm{tr}\{M(s_{\alpha_1}, {}^\sigma\pi^E) I_P(\sigma, \pi^E) I_P(\pi^E_{i\Lambda}, \phi)\} = \mathrm{tr}\, I_P(\pi_{i\Lambda}, f)$$

where $\pi$ is the representation $(\pi(\tau), \eta_3)$ of $M_1 = GL(2) \times GL(1)$. Since $\ell = 2$ there are exactly two $\pi$ which correspond to $\pi^E$, and these are $(\pi(\tau), \eta_3)$ and $(\pi(\tau), \zeta\eta_3)$.

Consider the function $m(P, A', \pi_{i\Lambda})$ of (b) and the function $m_E(P, A', {}^\sigma\pi^E_{i\Lambda})$ here. The group $A'$ is the same in both of these functions, which are logarithmic derivatives of functions defined with respect to $F$ and its extension $E$ of degree $\ell = 2$. Noting that $\pi(\tau) \otimes \xi \approx \pi(\tau)$ we have

$$\sum m(P, A', \pi_{i\Lambda}) = m_E(P, A', {}^\sigma\pi^E_{i\Lambda}) ,$$

where the sum is taken over the two $(= \ell)$ $\pi$ which correspond to $\pi^E$. Now $(8\pi)^{-1}$ is equal to $2$ $(= \ell)$ times $(16\pi)^{-1}$ and the $\pi$ thus obtained from our $\pi^E$ are the remaining terms of $I_\chi(f)$ in (b). We conclude that the sum of the terms in $I_\chi(f)$ of (b) indexed by $(\rho, \eta_3)$ with cuspidal $\rho$ of the form $\pi(\tau)$ is equal to $\ell$ times the sum of the terms of $I_\chi(\phi)$ considered here.

At this point we arrived to the conclusion that the sum of the terms from $\sum I_\chi(f)$ considered up to now is equal to $\ell$ times the sum of the terms from $\sum I_\chi(\phi)$ dealt with until now. However we have not studied as yet the last subcase of (c).

In passing we note that the comparison of the logarithmic derivatives $m$ and $m_E$ can be seen not only directly, as here, but also using the arguments proving Lemma 5.4 below.

4.2.1 (c) $p_\chi = \{P_0\}$, $A' = Z$

In the remaining subcase of (c) we have $A' = Z$. Here $P' = G$ and hence $A'$ reduces to a single point set. The set $\Delta_{P'}$ is empty and the operator $M(P,A',\pi_{i\Lambda})$ is the identity. The set $W^G(A_0)$ contains the two rotations $s = s_{\alpha_2\alpha_1}$ and $s_{\alpha_1\alpha_2}$. The value of the constant $c_s$ is 1/18. The corresponding contribution to $I_\chi(f)$ is

$$(18)^{-1}\sum_s \sum_\pi \mathrm{tr}\{M(s,\pi)I_P(\pi,f)\} . \qquad (8)$$

The sum over $\pi$ contains the representations $\pi$ of $P_0/N_0$ with $\pi = s\pi$ where $s\pi(g) = \pi(w_s^{-1}gw_s)$, that is if $\pi = (\eta_1,\eta_2,\eta_3)$ and $\eta_i$ are characters of $F^\times\backslash\mathbb{A}^\times$ then $\eta_1 = \eta_2 = \eta_3$ (on $\mathbb{A}^\times$, not only on $N\mathbb{A}_E^\times$).

The corresponding contribution to $I_\chi(\phi)$ is

$$(18)^{-1}\sum_s\sum_{\pi^E}\mathrm{tr}\{M(s,{}^\sigma\pi^E)I_P(\sigma,\pi^E)I_P(\pi^E,\phi)\} .$$

The sum over $\pi^E$ contains the representations of $P_0/N_0$ with ${}^\sigma\pi^E = s^{-1}\pi^E$, that is ${}^\sigma\pi^E = (\eta_3^E,\eta_1^E,\eta_2^E)$ if $s = s_{\alpha_1\alpha_2}$ and ${}^\sigma\pi^E = (\eta_2^E,\eta_3^E,\eta_1^E)$ if $s = s_{\alpha_2\alpha_1}$ where $\pi = (\eta_1^E,\eta_2^E,\eta_3^E)$.

Consider the terms associated with $\pi^E$ with $\pi^E = s\pi^E$. For such $\pi^E$ we have $\eta_1^E = \eta_2^E = \eta_3^E$ and ${}^\sigma\eta_i^E = \eta_i^E$ $(1 \le i \le 3)$. Hence there is a character $\eta$ of $\mathbb{A}^\times$ such that $\pi = (\eta,\eta,\eta)$ corresponds to $\pi^E$, and we have

$$\mathrm{tr}\{I_P(\sigma,\pi^E)I_P(\pi^E,\phi)\} = \mathrm{tr}\, I_P(\pi,f) .$$

There are exactly $\ell$ of the $\pi$ in $I_\chi(f)$ which correspond to $\pi^E$, obtained from $\zeta^i\eta$ $(0 \le i < \ell)$. The $M(s,\pi)$ and $M(s,\pi^E)$ are constants since they intertwine irreducible representations with themselves. They are readily seen to be independent of $F$ and $E$ (see below). We deduce that the contribution (8) to $I_\chi(f)$ is equal to $\ell$ times the subsum over $\pi^E$ with $s\pi^E = \pi^E$ in $I_\chi(\phi)$. We proved:

LEMMA 1. <u>We have</u>

$$\ell\sum I_\chi(\phi) - \sum I_\chi(f) = \frac{1}{6}\sum_s\sum_{\pi^E}\mathrm{tr}\{M(s,{}^\sigma\pi^E)I_P(\sigma,\pi^E)I_P(\pi^E,\phi)\} ,$$

<u>where</u> $s$ <u>is a rotation</u> ($w_{\alpha_1\alpha_2}$ <u>or</u> $w_{\alpha_2\alpha_1}$) <u>and</u> $\pi^E$ <u>is a triple</u> $((\eta^E, {}^{\sigma^2}\eta^E, {}^\sigma\eta^E)$ <u>or</u> $(\eta^E, {}^\sigma\eta^E, {}^{\sigma^2}\eta^E)$ , <u>respectively) such that</u> $\eta^E \ne {}^\sigma\eta^E$ <u>and</u> $\eta^E(Nx) = \omega_E(x)$ $(x$ <u>in</u> $\mathbb{A}_E^\times)$.

Here the coefficient 1/18 was replaced by 1/6 since we multiplied $\sum I_\chi(\phi)$ by $\ell$. The sum on the right is non-empty only when $\ell = 3$.

### 4.2.2 <u>Reformulation</u>

We shall discuss the right side above in more detail. We also change our notations so as to express our results in the way that they will be used in the next chapter. Let $S$ be the set of triples $\eta^E$ (instead of $\pi^E$) of unitary characters of $E^\times\backslash\mathbb{A}_E^\times$, such that for some rotation $w$ we have ${}^\sigma\eta^E = w^{-1}\eta^E$ and $w^{-1}\eta^E \neq \eta^E$. If $S$ is non-empty then $\ell = 3$. If $\eta^E$ is in $S$ then $I(\eta^E) = I_{P_0}(\eta^E)$ can be extended to a representation $I'(\eta^E)$ of $G \times G(\mathbb{A}_E)$ by

$$I'(\sigma,\eta^E) = M(w,{}^\sigma\eta^E)I(\sigma,\eta^E).$$

Indeed

$$I'(\sigma,\eta^E)I(g,\eta^E) = M(w,{}^\sigma\eta^E)I(\sigma,\eta^E)I(g,\eta^E)$$

$$= M(w,{}^\sigma\eta^E)I({}^\sigma g,{}^\sigma\eta^E)I(\sigma,\eta^E).$$

But $M(w,{}^\sigma\eta^E)$ intertwines $I({}^\sigma\eta^E)$ and $I(w{}^\sigma\eta^E) = I(\eta^E)$, hence the right side above is

$$= I({}^\sigma g,\eta^E)M(w,{}^\sigma\eta^E)I(\sigma,\eta^E) = I({}^\sigma g,\eta^E)I'(\sigma,\eta^E).$$

Moreover

$$(I'(\sigma,\eta^E))^3 = M(w,{}^\sigma\eta^E)M(w,{}^{\sigma^2}\eta^E)I(\sigma,\eta^E)I(\sigma,\eta^E)I'(\sigma,\eta^E)$$

$$= M(w,{}^\sigma\eta^E)M(w,{}^{\sigma^2}\eta^E)M(w,\eta^E)I(\sigma,\eta^E)I(\sigma,{}^{\sigma^2}\eta^E)I(\sigma,{}^\sigma\eta^E)$$

$$= M(w,{}^\sigma\eta^E)M(w,{}^{\sigma^2}\eta^E)M(w,\eta^E) = M(w,w^2\eta^E)M(w,w\eta^E)M(w,\eta^E).$$

From the theory of Eisenstein series (e.g. [1], I, (iii) on p. 927, $\ell$. 10) we have

$$M(w_1w_2,\eta^E) = M(w_1,w_2\eta^E)M(w_2,\eta^E). \tag{9}$$

Hence the expression $(I'(\sigma,\eta^E))^3$ is equal to

$$= M(w,w^2\eta^E)M(w^2,\eta^E) = M(w^3,\eta^E) = M(1,\eta^E) = 1.$$

The representations $I'(\eta^E)$ and $I'(w\eta^E)$ are equivalent since

$$M(w,\eta^E)I(g,\eta^E)M(w,\eta^E)^{-1} = I(g,w\eta^E)$$

and

$$M(w,\eta^E)M(w,{}^{\sigma}\eta^E)I(\sigma,\eta^E)M(w,\eta^E)^{-1} = M(w^2,w^2\eta^E)I(\sigma,\eta^E)M(w^2,w\eta^E)$$

(here we used (9) twice),

$$= M(w^2,w^2\eta^E)M(w^2,\eta^E)I(\sigma,w\eta^E) = M(w,{}^{\sigma}(w\eta^E))I(\sigma,w\eta^E)$$

$$= I'(\sigma,w\eta^E).$$

In fact $I'(\eta^E)$ is equivalent to each $I'(w\eta^E)$ where $w$ is any element in the Weyl group $W$ of $A_0$ in $G$. It suffices to show this for the reflection $w_{\alpha_1}$ since $W$ is generated by $w_{\alpha_1}$ and any rotation. We may also assume that $w = w_{\alpha_1\alpha_2}$. Then

$$M(w_{\alpha_1},\eta^E)I(g,\eta^E)M(w_{\alpha_1},\eta^E)^{-1} = I(g,w_{\alpha_1}\eta^E) ,$$

and

$$M(w_{\alpha_1}, \eta^E)M(w_{\alpha_1\alpha_2}, w_{\alpha_2\alpha_1}\eta^E)I(\sigma,\eta^E)M(w_{\alpha_1}, w_{\alpha_1}\eta^E)$$

$$= M(w_{\alpha_2}, w_{\alpha_2\alpha_1}\eta^E)I(\sigma,\eta^E)M(w_{\alpha_1}, w_{\alpha_1}\eta^E)$$

$$= M(w_{\alpha_2}, w_{\alpha_2\alpha_1}\eta^E)M(w_{\alpha_1}, w_{\alpha_1}w_{\alpha_2\alpha_1}\eta^E)I(\sigma, w_{\alpha_1}\eta^E)$$

$$= M(w_{\alpha_2\alpha_1}, w_{\alpha_1}w_{\alpha_2\alpha_1}\eta^E)I(\sigma, w_{\alpha_1}\eta^E)$$

$$= M(w_{\alpha_2\alpha_1}, {}^{\sigma}(w_{\alpha_1}\eta^E))I(\sigma, w_{\alpha_1}\eta^E)$$

which is $I'(\sigma, w_{\alpha_1}\eta^E)$ since ${}^{\sigma}(w_{\alpha_1}\eta^E) = w^{-1}w_{\alpha_1}\eta^E$ with $w = w_{\alpha_2\alpha_1}$.

Since for every $w$ in $W$ the involution $\eta^E \longrightarrow w\eta^E$ has no fixed points on $S$ the expression $|W|^{-1} \oplus I'(w\eta^E)$ (where the sum is taken over all $w$ in $W$) is a well-defined (up to equivalence) representation of $G \times G(\mathbb{A}_E)$. Since the number of elements $|W|$ of $W$ is 6 we can deduce from Lemma 1 and the displayed identity on the first page of the chapter the following

LEMMA 2. *We have*

$$\ell \operatorname{tr} r(\phi') - \operatorname{tr} r(f) + \sum_{\eta^E} \operatorname{tr}\{I'(\sigma,\eta^E)I(\eta^E,\phi)\} = \ell \sum I_0(\phi) - \sum I_0(f)$$

*where the sum on the left is* 0 *unless* $\ell = 3$, *and then it is taken over all unordered triples obtained by taking all conjugates of a character* $\mu^E$ *of* $E^\times\backslash\mathbb{A}_E^\times$ *with* ${}^{\sigma}\mu^E \neq \mu^E$ *and* $\mu^E(Nx) = \omega_E(x)$ $(x$ in $\mathbb{A}_E^\times)$.

<u>The extension</u> $I'(\eta^E)$ <u>of</u> $I(\eta^E)$ <u>to</u> $G \times G(\mathbb{A}_E)$ <u>is defined up to equivalence by</u>

$$I'(\sigma,\eta^E) = |W|^{-1} \otimes R(sws^{-1}, (s\eta^E))I(\sigma,s\eta^E);$$

<u>the sum is taken over all</u> $s$ <u>in</u> $W$ <u>and</u> $w$ <u>is defined by</u> ${}^{\sigma}\eta^E = w^{-1}\eta^E$.

The statement is clear if instead of the normalized intertwining operator $R$ we had the operator $M$. Suppose $\eta^E$ lies in $S$ and ${}^{\sigma}\eta^E = w^{-1}\eta^E$ with $w = w_{\alpha_1\alpha_2}$. By (9) we have

$$M(w_{\alpha_1\alpha_2},\eta^E) = M(w_{\alpha_1},w_{\alpha_2}\eta^E)M(w_{\alpha_2},\eta^E).$$

Since $\eta^E = (\mu^E,{}^{\sigma^2}\mu^E,{}^{\sigma}\mu^E)$ for some $\mu^E$ the results of the theory for $GL(2)$ imply that the normalizing factor for the first operator on the right is $L(1,{}^{\sigma}\mu^E/\mu^E)/L(1,\mu^E/{}^{\sigma}\mu^E)$, and for the second it is $L(1,{}^{\sigma}\mu^E/{}^{\sigma^2}\mu^E)/L(1,{}^{\sigma^2}\mu^E/{}^{\sigma}\mu^E)$. Since $R(w_{\alpha_1},w_{\alpha_2}\eta^E)R(w_{\alpha_2},\eta^E)$ satisfies the requirements of [3], §7, it is the normalized intertwining operator $R(w_{\alpha_1\alpha_2},\eta^E)$ and we obtain

$$M(w_{\alpha_1\alpha_2},\eta^E) = \frac{L(1,{}^{\sigma}\mu^E/\mu^E)}{L(1,{}^{\sigma^2}\mu^E/{}^{\sigma}\mu^E)}\,\frac{L(1,{}^{\sigma}\mu^E/{}^{\sigma^2}\mu^E)}{L(1,\mu^E/{}^{\sigma}\mu^E)}\;R(w_{\alpha_1\alpha_2},\eta^E).$$

Since $L(s,\chi) = L(s,{}^{\sigma}\chi)$ for all characters $\chi$ of $E^\times\backslash\mathbb{A}_E^\times$ the right side here is $R(w_{\alpha_1\alpha_2},\eta^E)$ and the lemma follows.

## 4.3.1 <u>The Hecke algebra</u>

We still have to prove that both sides of the identity of Lemma 2 are 0. This will be done by expressing both sides in terms of the Satake transform of some component $f_{v_0}$ of $f$. The left side will be written in the desired form in the remains of this chapter, and the discussion of the right side will be carried out in the next chapter.

Let $v$ be a non-archimedean place. The isomorphism classes of admissible irreducible unramified representations of $G(F_v)$ are classified by $A_0(\mathbb{C})/W$. For each $z = \mathrm{diag}(s,t,r)$ in $A_0(\mathbb{C})/W$ there is an irreducible subquotient $\pi(\eta_z)$ of the induced representation $I_{P_0}(\eta_z)$ of $G$. Here $\eta_z$ is the unramified quasi-character of $P_0$ whose value at a matrix with diagonal $(\mathrm{diag}(\tilde{\omega}^{m_1},\tilde{\omega}^{m_2},\tilde{\omega}^{m_3})$ is $s^{m_1}t^{m_2}r^{m_3}$. We shall consider only those representations which transform under $NE_v^\times$ by a character $\omega = \omega_v$, hence we have to consider only the $z$ with $str = \omega(\tilde{\omega})$ if $v$ splits or ramifies in $E$ and with $str = \omega(\tilde{\omega}^\ell)$ if $v$ stays prime. Here $\tilde{\omega} = \tilde{\omega}_v$ is the local uniformizing parameter of $F_v$.

The representation $\pi(\eta_{\bar{z}})$ is equivalent to the representation complex conjugate to $\pi(\eta_z)$, and $\pi(\eta_{z^{-1}})$ is equivalent to the representation contragredient to $\pi(\eta_z)$. Hence $\pi(\eta_{\bar{z}})$ is equivalent to $\pi(\eta_{z^{-1}})$ if $\pi(\eta_z)$ is unitary. Moreover, if $\pi(\eta_z)$ is unitary then the absolute values of $s,t$ and $r$ in $\mathbb{C}$ are bounded between $|\tilde{\omega}|$ and $|\tilde{\omega}|^{-1}$ ([5a],pp. 81-2).

Let $X$ be the set of $z$ in $A_0(\mathbb{C})/W$ with $\bar{z} \equiv z^{-1} \bmod W$ and such that the absolute values of $s,t$ and $r$ lie between $|\tilde{\omega}|$ and $|\tilde{\omega}|^{-1}$. It is a compact Hausdorff space.

The Hecke algebra $H_v$ of spherical functions on $G(F_v)$ which are compactly supported modulo $NE_v$ and transform under $NE_v^\times$ by $\omega^{-1}$, is isomorphic to the algebra of finite Laurent series in $z$ on $A_0(\mathbb{C})/W$. The isomorphism is given by mapping $f_v$ in the Hecke algebra to its Satake transform $f_v^\wedge$ whose value $f_v^\wedge(z) = f_v(s,t)$ at $z = (s,t,r)$ is given by

$$f_v^\wedge(s,t) = f_v^\wedge(\pi(\eta_z))$$

$$= \operatorname{tr} \pi(\eta_z, f_v) = \begin{cases} \sum F_v(m,n)s^m t^n, & v \text{ splits or ramifies} \\ \sum F_v(\ell m, \ell n)s^{\ell m} t^{\ell n}, & v \text{ stays prime} \end{cases} .$$

Here we assumed that $f_v$ is obtained from some $\phi_v$ by the map $\phi_v \longrightarrow f_v$, and write $F_v(m,n)$ for the orbital integral (with the $\Delta$-factor) of $f_v$ at any regular element in $A_0(F_v)$ whose eigenvalues have the valuations $|\tilde{\omega}^m|$, $|\tilde{\omega}^n|$ and 1. The sums are over pairs $(m,n)$ of integers, and they reduce to finite sums by the basic properties of $F(h,f_v)$. In the last equality a constant depending only on the choice of measures was suppressed since it will not affect our calculations.

The algebra obtained by restricting these Laurent series to $X$ separates points of the compact Hausdorff space $X$, contains the constant functions, and also the complex conjugate of each of its elements. By virtue of the Stone-Weierstrass theorem this algebra is dense in the algebra $C(X)$ of continuous complex valued functions on $X$.

### 4.3.2 The discrete series

After these preliminaries we can return to the discussion of the left side of the identity of Lemma 2. Note that multiplicity one theorem for $G(\mathbb{A}_E)$ on $L_0^2(\omega_E)$ is valid, and a constituent $\pi^E$ (of $L_0^2(\omega_E)$), acting on $V(\pi^E)$, is irreducible if and only if ${}^{\sigma}\pi^E$ is. Since $\pi^E(\phi')$ maps $V(\pi^E)$ to $V({}^{\sigma}\pi^E)$ it follows that if ${}^{\sigma}\pi^E$ is not equivalent to $\pi^E$, that is $V({}^{\sigma}\pi^E) \neq V(\pi^E)$, then the trace of $r(\phi')$ on the direct sum over $j(0 \leq j < \ell)$ of $V({}^{\sigma^i}\pi^E)$ must be 0. But if $V(\pi^E) = V({}^{\sigma}\pi^E)$ then the semi-direct product $G \times G(\mathbb{A}_E)$ acts on $V(\pi^E)$; the extended representation is denoted again by $\pi^E$. If $\pi^E = \otimes\pi_v^E$ then ${}^{\sigma}\pi_v^E \simeq \pi_v^E$ for each $v$ so that we can extend $\pi_v^E$ to a representation of $G \times G(E_v)$ which is determined up to a character of $G$.

Let $V$ be a fixed finite set of places containing the infinite ones and the finite places which ramify in E. Consider $\phi$ for which $\phi_v$, and hence $f_v = b(\phi_v)$, is spherical outside V. If $\pi_v^E$ (v outside V) is unramified we may suppose that $\pi_v^E(\sigma)$ fixes the $K(E_v)$-invariant vector. Thus

$$\operatorname{tr} \pi_v^E(\phi_v') = f_v^{\wedge}(z(\pi_v^E)) \qquad (v \text{ outside } V).$$

Here we define $\eta^E$ by $\pi_v^E = \pi(\eta^E)$ and let $\eta$ be any unramified character of $NA_0(F_v)$ with $\eta(Nx) = \eta^E(x)$; the $z = z(\pi_v^E)$ is defined by $\eta_z = \eta$.

Let $v_0$ be a fixed place outside V; in the next chapter we shall assume that $v_0$ splits in E. For any $z$ in $A_0(\mathbb{C})/W$ consider

$$\beta(z) = \sum_{\pi^E} \Pi_{v\notin V, v\neq v_0} f_v^{\wedge}(z(\pi_v^E))\cdot \Pi_{v \text{ in } V} \operatorname{tr} \pi_v^E(\phi_v')$$

where the sum is taken over all irreducible $\pi^E = \otimes\pi_v^E$ which are equivalent to ${}^{\sigma}\pi^E$, for which $\pi_v^E$ is unramified outside V and such that the component $\pi_{v_0}^E$ at $v_0$ satisfies $z(\pi_{v_0}^E) = z$ if $v_0$ splits and $z(\pi_{v_0}^E)^{\ell} = z^{\ell}$ if $v_0$

stays prime in E. Since $r(\phi')$ is of trace class the sums and products which define $\beta(z)$ are absolutely convergent for all $z$, the set of $z$ with $\beta(z) \neq 0$ is countable and denoted by $\{z_k;\ k \geq 0\}$, and we have

$$\operatorname{tr} r(\phi') = \sum_{k\geq 0}\beta(z_k)f^{\wedge}_{v_0}(z_k),$$

where the sum is absolutely convergent and taken over distinct $z_k$. Here $z$ and $z'$ are said to be distinct if $z \not\equiv z' \pmod W$ when $v_0$ splits in $E$ and if $z^{\ell} \not\equiv z'^{\ell} \pmod W$ if $v$ stays prime in $E$.

Similarly we can express $\operatorname{tr} r(f)$ as the absolutely convergent sum

$$\operatorname{tr} r(f) = \sum_{k\geq 0} \beta'(z'_k)f^{\wedge}_{v_0}(z'_k)$$

where $\{z'_k;\ k \geq 0\}$ is a sequence of distinct (in the above sense) elements of $A_0(\mathbb{C})/W$, and $\beta'(z'_k)$ are complex numbers.

4.3.3 A sum

To deal with the sum which occurs on the left of the identity in Lemma 2 we write $R(w,{}^{\sigma}\eta^{E})$ as a product $\otimes R(w,{}^{\sigma}\eta_v^{E})$. The local factor $R(w,{}^{\sigma}\eta_v^{E})$ fixes the $K(E_v)$-invariant vectors whenever $\eta_v^{E}$ is unramified; in fact this property of the normalized intertwining operators follows from the theory of GL(2), as we saw in the proof of Lemma 2.

For $v$ outside $V$ our $\phi_v$ is spherical and so $\operatorname{tr} I_{P_0}(\eta_v^{E},\phi'_v) = \operatorname{tr} I_{P_0}(\sigma)I_{P_0}(\eta_v^{E},\phi_v)$ is 0 unless $\eta_v^{E}$ is unramified, in which case it is $f^{\wedge}_v(z(\eta_v^{E}))$. Here $z = z(\eta_v^{E})$ is determined by the relation $\eta = \eta_z$ where $\eta = \eta_v$ is the unramified character of $NA_0(E_v)$ defined by $\eta_v(Nx) = \eta_v^{E}(x)$.

We can write this contribution to the trace formula as the absolutely convergent sum

$$\sum_{k\geq 0} \beta''(z_k'')f_{v_0}^{\wedge}(z_k'')$$

where $z_k''$ are distinct and $\beta''(z_k'')$ are complex numbers defined by

$$\beta''(z) = \sum_{\eta^E} \Pi_{v \text{ in } V} \operatorname{tr}\{R(w,{}^{\sigma}\eta_v^E)I_{P_0}(\sigma,\eta_v^E)I_{P_0}(\eta_v^E,\phi_v)\}$$

$$\cdot\ \Pi_{v\notin V, v\neq v_0} f_v^{\wedge}(z(\eta_v^E)) ,$$

where the sum is taken over the $\eta^E$ with ${}^{\sigma}\eta^E = w\eta^E \neq \eta^E$ as in Lemma 2 such that $\eta_v^E$ is unramified for $v$ outside $V$ and $z(\eta_{v_0}^E) = z$. We wrote $z_k''$ for the countable set of $z$ for which $\beta''(z)$ is non-zero.

We sum up this discussion with the following

LEMMA 3. _The left side of the identity in Lemma_ 2 _is equal to the absolutely convergent sum_

$$\sum_{k\geq 0} \beta_k f_{v_0}^{\wedge}(s_k,t_k) ,$$

_with elements_ $z_k = (s_k,t_k,r_k)$ _of_ $A_0(\mathbb{C})/W$ _which are distinct in the above sense. In fact the_ $z_k$ _lies in the compact subset_ $X$ _of_ $A_0(\mathbb{C})/W$.

The last sentence follows from the fact that all $\pi_{v_0}, \pi_{v_0}^E$ and $\eta_{v_0}^E$ which contribute to this sum are unitary.

## 5. EQUALITY OF TRACES

### 5.1.1 Elliptic terms

Our aim in this chapter is to prove that both sides of Lemma 4.2 are 0 for $\phi = \otimes\phi_v$ and $f = \otimes f_v$ with $\phi_v \longrightarrow f_v$. We shall also put the restriction of the second paragraph of 3.4.1; it does not restrict the generality of the results.

In Lemma 4.3 the left side was put in a form ready to be used, and we have to write the right side in a compatible way. The right side will be expressed as an integral in $f^{\wedge}_{v_0}$, and as in [12] it will be shown that such an integral cannot be equal to the discrete sum of Lemma 4.3 unless both sides are zero. In chapters 2 and 3 we studied the asymptotic behavior of $I(h,f)$ and $I(N\gamma,\phi)$ and showed in Lemma 2.8 that the summation formula can be applied to such non-smooth functions with that asymptotic behavior. It is here that the summation formula will be used to rewrite the sums of $I(h,f)$ and $I(N\gamma,\phi)$.

The main part of the trace formula is the sum over a set of representatives for the conjugacy classes of elliptic elements in $G(F)$ modulo $NZ(F)$ of the terms displayed in Lemma 2.1 together with the sum

$$\ell \sum_{h \text{ in } Z(F)/NZ(E)} |G(F)Z(\mathbb{A})\backslash G(\mathbb{A})| f(h)$$

of (3) Lemma 2.4. The main part of the twisted trace formula is the sum over a set of representatives $\gamma$ for the $\sigma$-conjugacy classes of elements in $G(E)$ modulo $Z(E)$ such that $N\gamma$ is either elliptic or a scalar in $F^{\times}$ but not in $NE^{\times}$ of the $J(N\gamma,\phi)$ of Lemma 3.1, together with the contribution

$$|G(F)Z(\mathbb{A})\backslash G(\mathbb{A})| \int_{Z(\mathbb{A}_E)G(\mathbb{A})\backslash G(\mathbb{A}_E)} \phi(g^{-\sigma}g)\,dg$$

((3) of Lemma 3.8).

First we note that if $h$ is an index in the first sum then the corresponding term vanishes (by our choice of $f$ ) unless $h$ is a local norm everywhere and hence a global norm. Without loss of generality we choose representatives so that $h = N\gamma$ . We claim that $\varepsilon(h) = \varepsilon(\gamma)$ . Indeed if $g^{-1}hg = zh$ for $g$ in $G(F)$ and $z = Nx \neq 1$ with $x$ in $Z(E)$ then $N(g^{-1}\gamma g) = N(x\gamma)$ and hence $x\gamma = \delta^{-\sigma}g^{-1}\gamma g\delta = (g\delta)^{-\sigma}\gamma g\delta$ for some $\delta$ in $G(E)$ .

Since on discrete sets we choose the measure which assigns 1 to each point we have

$$|Z_E(\mathbb{A})G_h(F)\backslash G_h(\mathbb{A})| = \ell|Z(\mathbb{A})G_h(F)\backslash G_h(\mathbb{A})| .$$

Since $G_\gamma^\sigma$ is a form of $G_h$ we also have

$$|Z(\mathbb{A})G_h(F)\backslash G_h(\mathbb{A})| = |Z(\mathbb{A})G_\gamma^\sigma(E)\backslash G_\gamma^\sigma(\mathbb{A}_E)| ,$$

a fact which follows from Lemma 2.9 as in [8]. Since $\Sigma I_0(\phi)$ has to be multiplied by $\ell$ , on expressing the global integrals as products of local ones, we can deduce from the local theory of orbital integrals (Lemmas 1.3 and 1.4) that the contributions from $\ell\Sigma I_0(\phi) - \Sigma I_0(f)$ under consideration cancel each other.

If $Z_E(\mathbb{A})G(F)\backslash G(\mathbb{A})$ was compact the trace formulae would have no terms other than those considered here, and we could deduce that both sides of the equality in Lemma 4.2 are 0 . But this is not the case and we have to work much harder, and indirectly, to deal with the other terms there.

5.1.2. Back to GL(2) .

For the comparison we shall need to relate the distribution $I_1(h,f_v)$ on $GL(2,F_v)$ , which is the inverse Fourier transform of $B(f_v,\eta)$ from [12], p. 177, line 5, to the distribution $I_1(h,\phi_v)$ on $GL(2,E_v)$ , which is the inverse

Fourier transform of $B(\phi_v,\eta_v)$ from [12], p. 197, line 5. In fact we shall assume that the distributions $B_1$ of [12] were defined by means of the corrected distributions $J_1(h,f_v)$ and $J_1(h,\phi_v)$ of 2.3.1 and 3.2.1 rather than using $A_3$ of [12], but the reader may note that the following result is valid with both definitions by the properties of $A_2$ of [12] and the linearity of the weight factor in the case of $GL(2)$.

Let $v_0$ be a fixed finite place of $F$ which splits in $E$.

LEMMA 1. <u>For every</u> $h$ <u>in</u> $NA(E)$ <u>and</u> $\phi = \otimes\phi_v$, $f = \otimes f_v$ <u>with</u> $\phi_v \longrightarrow f_v$ <u>for all</u> $v$ (<u>but</u> $v_0$) <u>we have</u>

$$\sum_{v\neq v_0} (I_1(h,\phi_v)-\ell I_1(h,f_v))\prod_{w\neq v,v_0} F(h,f_w) = 0 .$$

The notations are those of $GL(2)$. The sum does not include any place $v$ of $F$ which splits or is unramified in $E$ where $\phi_v$ (hence also $f_v$) is spherical. The lemma will be deduced from the result of [12], pp. 202-211, stating that

$$\sum_h \sum_v (I_1(h,\phi_v)-\ell I_1(h,f_v))\prod_{w\neq v} F(h,f_w) = 0$$

for all $\phi$ and $f$ as in the lemma. The sum over $h$ is taken over $NA(E)/NZ(E)$ since $F(h,f_w) \neq 0$ only if $h$ lies in $NA(E_w)$.

Given any $h_0$ in $NA(E)/NZ(E)$ we shall consider $f_{v_0}$ such that $F(h_0,f_{v_0})$ is non-zero. The function $f_{v_0}$ is left-invariant by some small (compact) subgroup $K'$ of $K(F_{v_0})$, and we shall consider the function $\phi_{v_0}$ defined before Lemma 1.13, with $f_1 = f_{v_0}$ and $f_i$ $(2\leq i\leq \ell)$ equal to the function which vanishes outside $Z(F_{v_0})K'$, whose value on $K'$ is $1/|K'|$ where $|K'|$ is the measure of $K'$. As noted after Lemma 1.16, that lemma is valid for this function and the operator $A = R^{-1}(\eta_{v_0})R'(\eta_{v_0})$ of [12]. This, together with Lemma 1.15 (or rather its statement which is valid for our non-spherical function $\phi_{v_0}$), implies that the sum over $v$ can be taken over a subset of the original set, with $v_0$ excluded. Hence there is always a factor $F(h,f_{v_0})$ in the product over $w$.

The sum over $h$ is finite, and is taken over a set which depends on the support of $f$ and $\phi$. Indeed if $F(h,\phi_w) \neq 0$ then the w-valuation of the quotient of the eigenvalues of $h$ lies between $C_w$ and $C_w^{-1}$, where $C_w \geq 1$ for all $w$ and $C_w = 1$ for almost all $w$. The product formula on $F^\times$, together with the fact that the sum over $v$ is finite, implies the assertion concerning the sum over $h$.

If we change $f_{v_0}$ and make $F(h,f_{v_0})$ have smaller support, the set over which the sum of $h$ is taken can become only smaller. In fact since the set of $h$ is discrete we may choose $f_{v_0}$ for which $F(h_0,f_{v_0})$ is non-zero but $F(h,f_{v_0}) = 0$ for all $h \neq h_0$ in the sum. For this $f_{v_0}$ the sum over $h$ has only one entry, the one corresponding to $h_0$, and since $F(h_0,f_{v_0}) \neq 0$ the lemma follows.

5.1.3. Excluding the place $v_0$.

For the proof of Proposition 5 below, we fix a finite place $v_0$ of $F$ which splits completely in $E$, regard the components $\phi_v$ and $f_v$ as fixed for $v \neq v_0$, and deal only with spherical $\phi_{v_0} \longrightarrow f_{v_0}$, where $\phi_{v_0} = (f_{v_0}, f^0_{v_0}, \ldots, f^0_{v_0})$.

The next part of the difference between $\sum I_0(f)$ and $\ell \sum I_0(\phi)$ to be considered is the difference between the sum over all $h$ in $NZ(E)\backslash A_0(F)$ of the values of $h$ of the two displayed functions in Lemma 2.6, and the corresponding sum over $h$ in $NZ(E)\backslash NA_0(E)$ of the values of the functions displayed in Lemma 3.10, multiplied by $\ell$.

Consider first the difference involving the second function in each of these lemmas. The function of Lemma 2.6 vanishes unless $h$ lies in $NA_0(E)$ since $F(h,f_w) = 0$ unless $h$ lies in $NA_0(E_w)$. Since $f_{v_0}$ and $\phi_{v_0}$ are spherical the properties of the distribution $I$ assure us that

$$I(h,f_{v_0}) = J(h,f_{v_0}) \ , \quad I(h,\phi_{v_0}) = J(h,\phi_{v_0}) \ .$$

Lemma 1.15 holds for our functions $\phi_{v_0}$ and $f_{v_0}$. Since

$$\log|u|_{E_v} = \log|N_{E_v/F_v}u|_{F_v} = \log|u^\ell|_{F_v} \qquad (u \text{ in } F_v^\times)$$

and the correction of $J(h,f_v)$ and $J(h,\phi_v)$ was made by means of elements in $F^\times$, Lemma 1.15 with $q = 2$ implies that

$$J(h,\phi_{v_0}) = \ell^2\, J(h,f_{v_0}) \ .$$

Since the function of Lemma 3.10 has to be multiplied by $\ell$, we deduce that in the difference under consideration we may assume that the sums over $v$ which define the functions are taken only over $v \neq v_0$.

Next we consider the difference of the sums of values of the first displayed functions in Lemma 2.6 and 3.10, and claim that the sums over $v_1$ and $v_2$ have to be taken only over those $v_i$ with $v_i \neq v_0$ $(i = 1,2$ ; as well as $v_1 = v_2$, as before). Indeed, if $v_1 = v_0$, we have to consider the difference between the sum over $h$ in $NZ(E)\backslash A_0(F)$ of

$$\ell^2 \sum_{v \neq v_0} I_1(h,f^K_{v_0N_1})I_1(h,f^K_{vN_1})\prod_{w\neq v,v_0} F(h,f_w) \ ,$$

and the sum over $h$ in $NZ(E)\backslash NA_0(E)$ of

$$\sum_{v\neq v_0} I_1(h,\phi^K_{v_0N_1})\, I_1(h,\phi^K_{vN_1}) \prod_{w\neq v,v_0} F(h,\phi_w)\ .$$

Since $NA_0(E_{v_0}) = A_0(F_{v_0})$ and $F(h,f_w) = 0$ unless $h$ lies in $NA_0(E_w)$, the first sum over $h$ has to be taken only over $NZ(E)\backslash NA_0(E)$. But $f_{v_0}, \phi_{v_0}$, and hence the functions $f^K_{v_0N_1}, \phi^K_{v_0N_1}$ on $GL(2)$, are spherical, and we have

$$I_1(h,f^K_{v_0N_1}) = J_1(h,f^K_{v_0N_1})\ ,\quad I_1(h,\phi^K_{v_0N_1}) = J_1(h,\phi^K_{v_0N_1})\ .$$

Lemma 1.15 applies with $q = 1$, and we deduce that

$$J_1(h,\phi^K_{v_0N_1}) = \ell J_1(h,f^K_{v_0N_1})\ .$$

It remains to consider

$$\sum_{v\neq v_0} (I_1(h,\phi^K_{vN_1}) - \ell I_1(h,f_{vN_1}))\, J_1(h,f^K_{v_0N_1}) \prod_{w\neq v,v_0} F(h,f_w)\ ,$$

since $F(h,f_w) = F(h,\phi_w)$ for all $w$ and $h$. This is the product by $J_1(h,f^K_{v_0N_1})$ of the left side of the expression of Lemma 1, and the assertion follows.

We conclude that in the study of the difference we may assume that the sums over $v,v_1,v_2$ in Lemmas 2.6 and 3.10 are taken over (finite) sets excluding $v_0$, and the products over $w$ always include the factor $F(h,f_{v_0})$ (or $F(h,\phi_{v_0})$).

5.1.4. <u>Split terms</u>.

Under the usual conditions (of 5.1.1 and 5.1.3) on $\phi$ and $f$, we have:

LEMMA 2. <u>There exists a bounded integrable function</u> $\beta(s,t)$ <u>on</u> $|s| = |t| = 1$ <u>in</u> $\mathbb{C}^{\times 2}$ <u>such that the difference considered in</u> 5.1.3 <u>is equal to</u>

$$\left(\frac{1}{2\pi i}\right)^2 \iint_{|s|=|t|=1} \beta(s,t) f^{\wedge}_{v_0}(s,t)\, d^{\times}s\, d^{\times}t\ ,$$

<u>where</u> $f^{\wedge}_{v_0}$ <u>is the transform of</u> $f_{v_0}$ <u>from</u> 4.3.1.

Proof. Let $D$ denote the set of quasi-characters of $A_0(\mathbb{A})$ which transform under $Z_E(\mathbb{A})$ by the character $\omega$. This set has a structure of a complex manifold of dimension 2 by the parametrization $\eta\alpha(s,t)$ ($s,t$ in $\mathbb{C}$) at a neighborhood of $\eta$ in $D$. Here $\alpha(s,t) = \otimes\alpha_v(s,t)$ is a quasicharacter of $A_0(\mathbb{A})$ ; the local factor $\alpha_v(s,t)$ is the unramified quasi-character of $A_0(F_v)$ whose value at $(a,b,c)$ is $|a/c|_v^s|b/c|_v^t$. Since the index of $Z_E(\mathbb{A})$ in $Z(\mathbb{A})$ is $\ell$ the dual measure on $D$ is $\ell^{-1}(2\pi)^{-2}dsdt$. The part from the trace formula in which we are interested is the sum over $h$ in $NZ(E)\backslash A_0(F)$ of

$$\ell\lambda_{-1}^2[\sum_{v_1\neq v_2}\prod_{i=1,2} I_1(h,f^K_{v_iN_1})\prod_{w\neq v_1,v_2} F(h,f_w) + \sum_v I(h,f_v)\prod_{w\neq v} F(h,f_w)] .$$

By virtue of Lemma 2.8 and the work of 2.7.1 and 2.7.2, the summation formula can be applied to this function of $h$, to the group $NZ(\mathbb{A}_E)\backslash A_0(\mathbb{A})$ and its subgroup $NZ(E)\backslash A_0(F)$. The Fourier transform has to be taken over the set $D^0$ of unitary characters. The relation between the global and local Tamagawa measures on $Z\backslash A_0$ lets us compute the Fourier transform locally and then divide by $\lambda_{-1}^2$, instead of calculating it globally. We shall denote the Fourier transform of a distribution $A$ by $A^\wedge$. For example the transform of $F(h,f_v)$ at $\eta_v$ is

$$F^\wedge(\eta_v,f_v) = \int_{NZ(E_v)\backslash NA_0(E_v)} F(h,f_v)\eta_v(h)dh$$
$$= \int_{A_0(E_v)^{1-\sigma}Z(E_v)\backslash A_0(E_v)} F(N\gamma,\phi_v)\eta_v^E(\gamma)d\gamma = F^\wedge(\eta_v^E,\phi_v) ,$$

where $\eta_v^E(\gamma) = \eta_v(N\gamma)$ ( $\gamma$ in $A_0(E_v)$ ), since $\phi_v \longrightarrow f_v$. We obtain

$$(\frac{1}{2\pi})^2\int_{D^0}\{\sum_{v_1\neq v_2}\prod_{i=1,2} I_1^\wedge(\eta_{v_i},f^K_{v_iN_1})\prod_{w\neq v_1,v_2} F^\wedge(\eta_v,f_v)$$
$$+ \sum_v I^\wedge(\eta_v,f_v)\prod_{w\neq v} F(\eta_v,f_v)\} dsdt .$$

An analogous discussion can be carried out for the part from the twisted trace formula under consideration. The set $D_E$ of quasi-characters of $Z(\mathbb{A}_E)A_0(\mathbb{A}_E)^{1-\sigma}\backslash A_0(\mathbb{A}_E)$ has a structure of a two-dimensional complex manifold with the usual parametrization, defined with respect to $E$. The dual measure is given by $\ell^2(2\pi)^{-2}dsdt$ since our use of the function $\hat{\tau}$ with respect to $E$ forced us to multiply the function of $\gamma$ by $\ell^{-2}$. Note that if $\eta^E$ in $D_E^0$ is invariant under $A_0(\mathbb{A}_E)^{1-\sigma}$ then ${}^\sigma\eta^E = \eta^E$. We consider the sum over $\gamma$ in $A_0(E)^{1-\sigma}Z(E)\backslash A_0(E)$ of

$$\ell^{-2}\lambda_{-1}^2[\sum_{v_1\neq v_2}\prod_{i=1,2} I_1(N\gamma,\phi^K_{v_iN_1})\prod_{w\neq v_1,v_2} F(N\gamma,\phi_w)$$

$$+\sum_v I(N\gamma,\phi_v)\prod_{w\neq v} F(N\gamma,\phi_w)]\ .$$

Lemma 2.8 and the comments of 3.4.1 and 3.4.2 show that the summation formula can be applied to this function of $\gamma$, the group $A_0(\mathbb{A}_E)^{1-\sigma}Z(\mathbb{A}_E)\backslash A_0(\mathbb{A}_E)$ and its subgroup $A_0(E)^{1-\sigma}Z(E)\backslash A_0(E)$. The Fourier transform has to be taken over the set $D_E^0$ of unitary characters. Dividing by $\lambda_{-1}^2$ and calculating the Fourier transform locally we get

$$(2\pi)^{-2}\int_{D_E^0}\{\sum_{v_1\neq v_2}\prod_{i=1,2} I_1^\wedge(\eta^E_{v_i},\phi^K_{v_iN_1})\prod_{w\neq v_1,v_2} F(\eta^E_w,\phi_w)$$

$$+\sum_v I^\wedge(\eta^E_v,\phi_v)\prod_{w\neq v} F^\wedge(\eta^E_w,\phi_w)\}dsdt\ .$$

Recalling the conclusion of 5.1.3 we deduce that the difference under consideration is equal to

$$(2\pi)^{-2}\int_{D_E^0} \hat{f_0}(t(\eta_{v_0}))\beta(\eta^E)dsdt \qquad (\eta_{v_0}(Nx)=\eta^E_{v_0}(x))\ .$$

Here $\beta(\eta^E)$ is a finite sum over $v$ of products indexed by $w \neq v_0$, almost all of whose factors are of the form $F^\wedge(\eta_w^E, \phi_w^0)$. Note that $F^\wedge(\eta_w^E, \phi_w^0) \neq 0$ only if $\eta_w^E$ is unramified. The integral is taken over the $\sigma$-invariant characters $\eta^E$ which are unramified outside a fixed set of places, and at $v_0$.

Let $\eta^E$ be a representative in each connected component of the domain of integration with $\eta_{v_0}(\tilde{\omega}_{v_0}) = (1,1,\omega(\tilde{\omega}_{v_0}))$. Write $\beta(s,t)$ for the sum over $\eta^E$ of $\beta(\eta^E\alpha(s,t))$. The integral becomes

$$(2\pi)^{-2} \int_{\mathbb{R}}\int_{\mathbb{R}} \beta(is,it) f_{v_0}^\wedge(|\tilde{\omega}|^{is}, |\tilde{\omega}|^{it}) ds dt .$$

Replacing $\beta(is,it)$ by the sum over all integers $j,k$ of

$$\beta\left(is + \frac{2\pi ij}{\log|\tilde{\omega}|},\ it + \frac{2\pi ik}{\log|\tilde{\omega}|}\right)$$

and passing to multiplicative notations, changing $|\tilde{\omega}|^{is}$ to $s$, $|\tilde{\omega}|^{it}$ to $t$, we obtain the expression of the lemma.

The lemma follows on noting that the integral over $D_E^0$ as well as the (finite) sum and products which define $\beta(\eta^E)$ are absolutely convergent, and that the algebra of Satake transforms $f_{v_0}^\wedge$ is dense in the algebra of continuous functions on $X$ (in particular, on $|s|=|t|=1$).

### 5.2.1. Quadratic terms

In this section we shall study the difference between two sums. They are:

(1) The sum over (i) $h$ in $NA_1(E)\backslash A_1(F)$ and (ii) over a set of representatives $h$ in $M_1(F)$ for the conjugacy classes of quadratic elements in $NA_1(E)\backslash M_1(F)$ of

$$\ell\, c_h\, \lambda_{-1} \sum_a \sum_v I(ah,f_v) \prod_{w \neq v} F(ah,f_w) ;$$

here the sum over $a$ is taken over a set of representatives in $NA_1(E)$ for $NZ(E)\backslash NA_1(E)$ (excluding the class of $a = 1$ if $h = 1$), and

$$c_h = |G_h(F)A_1(\mathbb{A})\backslash G_h(\mathbb{A}))|$$

if $h \neq 1$ and $c_1 = \frac{3}{2}c_h$ for any $h$ in $A_1(F) - Z(F)$. This combines the terms of Lemma 2.2 and (2) of Lemma 2.3. When $h = 1$, $I(ah,f_v)$ is defined by the local analogue of the integral of (2), Lemma 2.3 ($h$ there is our $a$); $F(ah,f_v)$ is defined by the same local integral, but with the weight factor omitted.

(2) The sum over (i) $\gamma = 1$, and (ii) a set of representatives $\gamma$ in $A_1(E)\backslash M_1(E)$ for the $\sigma$-conjugacy classes of the elements of $G(E)$ whose norms are quadratic in $G(F)$ or they have exactly two equal eigenvalues in $F^\times$ but not in $NE^\times$, of

$$c_{N\gamma}\lambda_{-1} \sum_a \sum_v I(N(a\gamma),\phi_v) \prod_{w \neq v} F(N(a\gamma),\phi_w) .$$

The sum over $a$ extends over $Z(E)A_1^{1-\sigma}(E)\backslash A_1(E)$, excluding the class of the identity if $\gamma = 1$. When $\gamma = 1$, $J(N(a\gamma),\phi_v)$ is defined by the local analogue of the integral of (2), Lemma 3.7; $F(N(a\gamma),\phi_w)$ is then defined by the same integral but with the weight factor omitted.

Standard facts [8] concerning comparison of measures on $GL(2)$ and its inner forms show that the sum of (2) is the same as the product by $\ell$ of the sum of the terms of Lemma 3.2 and (2) of Lemma 3.7.

Note that the sums over $h$ and $a$ in (1) are taken over sets isomorphic (by the norm map) to the sets over which the sums over $\gamma$ and $a$ in (2) are taken. To establish this note that $F(ah,f_w)$ vanishes unless $ah$ is a norm of an element in $G(F_w)$. Lemma 1.1 now implies that the sums over $a$ and $h$ in (1) can be taken only over $a$ and $h$ so that $ah$ is a norm, since $ah$ is a norm at all places but one if $ah$ contributes a non-zero term to (1). But $a$ is a norm, hence $h$ is a norm, as asserted.

To study our difference we may assume that the sums in both (1) and (2) are taken only over $v \neq v_0$. Indeed, consider the difference between these sums where instead of the sums over $v$ we take only the terms indexed by $v = v_0$. This last difference is equal to 0 since (1) $F(N\gamma,f_w) = F(N\gamma,\phi_w)$ for all $w$, (2) $\phi_{v_0}$ is spherical, hence (if $N\gamma$ has eigenvalues outside $F^\times$)

$$I(N(a\gamma),f_{v_0}) = J(N(a\gamma),f_{v_0}) , \quad I(N(a\gamma),\phi_{v_0}) = J(N(a\gamma),\phi_{v_0}) ,$$

(3) the correction of $J(N\gamma,\phi)$ when the eigenvalues of $N\gamma$ lie in $F^\times$ in 3.1.4, that of (2) in Lemma 2.3 and (2) of Lemma 3.7, were done in a consistent manner, hence by Lemma 1.15 we have

$$J(N(a\gamma)\,,\,\phi_{v_0}) = \ell J(N(a\gamma),f_{v_0}) .$$

For each $h$ we shall apply the summation formula to the function of $a$ given by the sum over $v$ in (1). Lemma 2.2 and considerations on the level of 2.7.1 applied to (2) of Lemma 2.3, together with Lemma 2.8, assure us that the formula can indeed be applied. The pair of groups will be $NZ(E)\backslash NA_1(E)$ and $NZ(\mathbb{A}_E)\backslash NA_1(\mathbb{A}_E)$.

In fact we shall assume that the sum over $a$ in (1), as well as the sum over $a$ in (2), both are extended to include the term $a = 1$ in the case that

$h = 1$ and $\gamma = 1$. These two terms will have to be subtracted later on. Note that for the groups over which we apply the summation formula it would have been enough not to correct the terms of Lemma 3.2 and the corresponding terms in (2) of Lemma 2.3. We chose the shortest way in which Lemma 2.3 could be expressed.

Consider the set of $D$ of quasi-characters of $NZ(\mathbb{A}_E)\backslash NA_1(\mathbb{A}_E)$. It has a structure of a complex manifold of dimension 1 through the parametrization $\eta\alpha(s)$ ($s$ in $\mathbb{C}$) in a neighborhood of $\eta$. Here $\alpha(s) = \Pi\alpha_v(s)$ and the local function $\alpha_v(s)$ is the unramified character whose value at $(a,a,b)$ is $|a/b|_v^s$. The factor $\ell$ appears in the expression of (1) since it is the index of $Z_E(\mathbb{A})$ in $Z(\mathbb{A})$. This is also the reason why the dual measure on $D$ is $(2\pi\ell)^{-1}ds$. Now in our case the Fourier transform has to be taken over the subset $D^0$ of unitary characters. The relation $d^\times a = (\lambda_{-1})^{-1} \otimes d^\times a_v$, between global and product of local Tamagawa measures on $A_1$, permits us to compute the Fourier transform locally, rather than globally, if we divide our function by $\lambda_{-1}$. If we denote the transforms of $F(ha,f_v)$ and $I(ha,f_v)$ with respect to $\eta_v$ by $F\hat{}(h,\eta_v,f_v)$ and $I\hat{}(h,\eta_v,f_v)$ the result is

$$\frac{c_h}{2\pi}\int_{D^0}\sum_{v\neq v_0} I\hat{}(h,\eta_v,f_v)\prod_{w\neq v} F\hat{}(h,\eta_w,f_w)ds .$$

For each $\gamma$ we shall apply the summation formula to the function of $a$ given by the sum over $v$ in (2), and the pair $A_1(E)^{1-\sigma}Z(E)\backslash A_1(E)$ and $A_1(\mathbb{A}_E)^{1-\sigma}Z(\mathbb{A}_E)\backslash A_1(\mathbb{A}_E)$. The formula can be applied by virtue of Lemma 3.3, considerations as in 3.4.2 applied to the term (2) of Lemma 3.7, and Lemma 2.8.

The set $D_E$ of quasi-characters of $Z(\mathbb{A}_E)A_1(\mathbb{A}_E)^{1-\sigma}\backslash A_1(\mathbb{A}_E)$ has a structure of a one-dimensional complex manifold with the usual parametrization, taken with respect to $E$. The dual measure on $D_E$ is $\frac{\ell}{2\pi}ds$, since the function of $\gamma$ had to be multiplied by $\ell^{-1}$. The Fourier transform has to be taken over the set $D_E^0$ of unitary characters. Using the usual relation between global and local Tamagawa measures on $A_1$ , and notations analogous to those used in the non-twisted case, we obtain

$$\frac{\ell c_{N\gamma}}{2\pi}\int_{D_E^0}\sum_{v\neq v_0} I^{\wedge}(N\gamma,\eta_v^E,\phi_v)\prod_{w\neq v} F^{\wedge}(N\gamma,\eta_w^E,\phi_w)ds .$$

It was noted above that each $h$ under consideration is a norm, hence of the form $N\gamma$ . For any $\eta_w$ with $\eta_w(Nx) = \eta_w^E(x)$ , that is $\eta_w \longrightarrow \eta_w^E$ , we have

$$F^{\wedge}(N\gamma,\eta_w,f_w) = \int_{NZ(E_w)\backslash NA_1(E_w)} F(aN\gamma,f_w)\eta_w(a)da$$

$$= \int_{A_1(E_w)^{1-\sigma}Z(E_w)\backslash A_1(E_w)} F(N(a\gamma),\phi_w)\eta_w^E(a)da = F^{\wedge}(N\gamma,\eta_w^E,\phi_w) .$$

Hence for each $\gamma$ of (2) we have the difference

$$\frac{1}{2\pi}\int_{D_E^0} F^{\wedge}(N\gamma,\eta_{v_0}^E,\phi_{v_0})\ \beta(N\gamma,\eta^E)ds ,$$

where

$$\beta(N\gamma,\eta^E) = \sum_v[\sum_{\eta\to\eta^E} I^{\wedge}(N\gamma,\eta_v,f_v) - \ell I^{\wedge}(N\gamma,\eta_v^E,\phi_v)]\prod_{w\neq v} F^{\wedge}(N\gamma,\eta_w^E,\phi_w) .$$

The integral is taken over the $\eta^E$ with unramified component $\eta_v^E$ at all $v$ where $\phi_v$ is spherical (and $v$ is finite unramified); this set of $v$ includes $v_0$ . Indeed $F^{\wedge}(N\gamma,\eta_v^E,\phi_v)$ vanishes if $\phi_v$ is spherical but $\eta_v^E$ is ramified. Note that for each $\eta^E$ in $D_E$ we have ${}^{\sigma}\eta^E = \eta^E$ .

Let us fix an $\eta^E$ in each connected component of such characters in $D_E^0$ with $\eta_{v_0}^E = (\eta_1,\eta_2)$ and $\eta_1(\tilde{\omega}_{v_0}) = 1$ , $\eta_2(\tilde{\omega}_{v_0}) = \omega(\tilde{\omega}_{v_0})$ . Since $F^{\wedge}(N\gamma,\eta_{v_0}^E\alpha_{v_0}(s),\phi_{v_0})$ does not depend on the connected component we denote it by $F^{\wedge}(N\gamma,e^{s\log|\tilde{\omega}|},\phi_{v_0})$ , and for each $\gamma$ we signify by $\beta'(N\gamma,s)$ the sum of $\beta(N\gamma,\eta^E)$ over a set of representatives $\eta^E$ as above. The integral is taken no longer over $s$ in $i\mathbb{R}$ but over $is$ with $0 \leq s \leq 2\pi/\log|\tilde{\omega}|$. We put

$$\beta(N\gamma, e^{i \log|\tilde{\omega}| s}) = \log|\tilde{\omega}| \sum_k \beta'(N\gamma, is + 2\pi ik/\log|\tilde{\omega}|),$$

and pass to multiplicative measures, changing $e^{i \log|\tilde{\omega}| s}$ to $s$. The integral becomes

$$\frac{1}{2\pi i} \int_{|s|=1} F^{\wedge}(N\gamma, s, \phi_{v_0}) \beta(N\gamma, s) d^{\times}s$$

5.2.2. Local quadratic tori

The difference of 5.2.1 is not the last integral of that subsection, but rather the sum over all $\gamma$ of 5.2.1 (2) of these integrals. Since all of the sums and products which make these integrals are absolutely convergent and the local Fourier transforms are absolutely integrable, we may take the sum over $\gamma$ inside the integral over $|s| = 1$. But $F^{\wedge}(N\gamma, s, \phi_{v_0})$ depends only on the image of $N\gamma$ in $M_1(F_{v_0})$ modulo $NA_1(E_{v_0}) = A_1(F_{v_0})$; hence we shall be able to add up the $\beta(N\gamma, s)$ over all global $\gamma$ with the given component at $F_{v_0}$. There are only a finite number of quadratic tori $T$ over $F_{v_0}$, and we shall deal with each case separately.

Suppose that $N\gamma$ is quadratic over $F$ and it splits over $F_{v_0}$. Since $N\gamma$ is chosen modulo $NA_1(E)$ we may assume that its eigenvalues $h_1, h_2, 1$ have the valuations $1, |\tilde{\omega}_{v_0}|^k, 1$ $(k \geq 0)$ at $v_0$. The Fourier transform $F^{\wedge}(N\gamma, s, f_{v_0}) =$ $= F^{\wedge}(N\gamma, s, \phi_{v_0})$ of $F(aN\gamma, f_{v_0})$ ($a$ in $A_1(F_{v_0})$) with respect to the unramified character $\eta_{v_0} = (\eta_1, \eta_2)$ of $A_1(F_{v_0})$ with $\eta_1(\tilde{\omega}^m_{v_0}) = s^m$ and $\eta_2(\tilde{\omega}^m_{v_0}) =$ $= (\omega(\tilde{\omega}_{v_0})/s)^m$ is

$$\sum_m F_{v_0}(m, m+k) s^m \quad (\text{integral } m).$$

The sum is finite since $f_{v_0}$ is compactly supported modulo the center. Further we have

$$\frac{1}{2\pi i}\int_{|t|=1} f^{\wedge}_{v_0}(s/t,t)t^{-k}d^{\times}t = \frac{1}{2\pi i}\int_{|t|=1} \sum_{m,n} F_{v_0}(m,n)s^m t^{n-m} t^{-k} d^{\times}t .$$

$$= (2\pi i)^{-1}\int_{|t|=1} \sum F_{v_0}(m,m+n)s^m t^n t^{-k} d^{\times}t = \sum F_{v_0}(m,m+k)s^m .$$

We write $k(\gamma)$ for $k$ to emphasize the dependence on $\gamma$ .

Let $\beta(k,s)$ be the sum over $\gamma$ such that $N\gamma$ is quadratic over $F$ , it splits over $F_{v_0}$ and $k(\gamma) = k$ . The corresponding part of the integrand of our integral over $|s| = 1$ is therefore

$$(2\pi i)^{-1}\int_{|t|=1} f^{\wedge}_{v_0}(s/t,t)(\sum_k \beta(k,s)t^{-k})d^{\times}t ,$$

where the sum is absolutely convergent.

Suppose the $N\gamma$ is quadratic over $F_{v_0}$ . Since $N\gamma$ is chosen in $M_1(F_{v_0})$ modulo $NA_1(F_{v_0})$ we may assume that its eigenvalues are $h_1,h_2,1$ , and that $|h_1h_2|_{v_0} = |\tilde{\omega}|^j_{v_0}$ with $j=0$ or $1$ . Let $h$ be an element of $GL(2,F)$ with eigenvalues $h_1,h_2$ . The Fourier transform $F^{\wedge}(N\gamma,s,f_{v_0})$ of $F(aN\gamma,f_{v_0})$ with respect to the above $\eta_{v_0}$ is

$$\sum_m F(\mathrm{diag}(\tilde{\omega}^m h,1),f_{v_0})s^m .$$

The coefficients here are the orbital integrals of $f_{v_0}$ at the $G(F_{v_0})$ element $\mathrm{diag}(\tilde{\omega}^m h,1)$ multiplied by the factor

$$\Delta\left(\begin{pmatrix} \tilde{\omega}^m h & 0 \\ 0 & 1 \end{pmatrix}\right) = \left| \frac{\tilde{\omega}^m h_1 - 1}{\tilde{\omega}^m h_1} \; \frac{\tilde{\omega}^m h_2 - 1}{\tilde{\omega}^m h_2} \; \frac{\tilde{\omega}(h_1-h_2)}{1} \right|_{v_0} .$$

A usual change of variables showsus that these coefficients are equal to

$$|\tilde{\omega}^m|_{v_0} |h_1 h_2|_{v_0}^{1/2} F(\tilde{\omega}^m h, f^K_{v_0 N_1}) ,$$

where, as usual,

$$f^K_{v_0 N_1}(g) = \iint f_{v_0}(k^{-1} gnk)\, dn dk$$

( $g$ in $M_1(F_{v_0})$ , $k$ in $M_1 \cap K\backslash K$ , $n$ in $N(F_{v_0})$ ). Here $f^K_{v_0 N_1}$ is viewed as a spherical function on $GL(2, F_{v_0})$ and $F$ is its orbital integral at $\tilde{\omega}^m h$ (in $GL(2, F_{v_0})$), multiplied, as usual, by

$$\Delta_1(\tilde{\omega}^m h) = |(\tilde{\omega}^m(h_1 - h_2))^2 / \tilde{\omega}^{2m} h_1 h_2|_{v_0}^{1/2} = |(h_1 - h_2)^2 / h_1 h_2|_{v_0}^{1/2} = \Delta_1(h) .$$

The above reduction makes it possible for us to apply some results on spherical functions over $GL(2)$ . Indeed if $j = 0$ then Lemmas 5.6 and 5.8 of [12] imply that

$$F(\tilde{\omega}^m h, f^K_{v_0 N_1}) = c_1 F(\begin{pmatrix} a & 0 \\ 0 & b \end{pmatrix}, f^K_{v_0 N_1}) - c_2 \Delta_1(h) f^K_{v_0 N_1}(\tilde{\omega}^m) . \qquad (*)$$

Here $a$ and $b$ are any distinct elements with valuations $|\tilde{\omega}^m|_{v_0}$ . The constants $c_1, c_2$ as well as $c_3$ below are independent of $h$ and $f^K_{v_0 N_1}$ ; their values are given explicitly in [12], but they are not important for us. A change of variables shows that

$$|\tilde{\omega}^m|_{v_0} F(\begin{pmatrix} a & 0 \\ 0 & b \end{pmatrix}, f^K_{v_0 N_1}) = F(\mathrm{diag}(a,b,1), f_{v_0}) = F_{v_0}(m,m) .$$

Denote by $\beta_0(s)$ the sum of $c_1 \beta(N\gamma, s)$ over the $\gamma$ such that $N\gamma$ is quadratic over $F_{v_0}$ and for which $j = 0$ . The corresponding part of our integrand can now be put in the form

$$(2\pi i)^{-1}\int_{|t|=1} f^{\wedge}_{v_0}(s/t,t)\beta_0(s)d^{\times}t\ .$$

When $j = 1$ the place $v_0$ is ramified in the extension generated by $h$ over $F_{v_0}$. The proofs of Lemmas 5.7 and 5.1 in [12] imply that

$$F(\tilde{\omega}^m h, f^K_{v_0N_1}) = c_3\Delta_1(h)\left[F\left(\begin{pmatrix}\tilde{\omega}^{m+1} & 0\\ 0 & \tilde{\omega}^m\end{pmatrix}, f^K_{v_0N_1}\right) - |\tilde{\omega}|^{1/2} f^K_{v_0N_1}\left(\begin{pmatrix}\tilde{\omega}^{m+1} & 0\\ 0 & \tilde{\omega}^m\end{pmatrix}\right)\right]. \quad (**)$$

Changing variables, noting that $|h_1h_2|_{v_0} = |\tilde{\omega}|_{v_0}$ and that $|\tilde{\omega}-1|_{v_0} = 1$ we deduce that

$$|\tilde{\omega}|^m_{v_0}|h_1h_2|^{1/2}_{v_0}F\left(\begin{pmatrix}\tilde{\omega}^{m+1} & 0\\ 0 & \tilde{\omega}^m\end{pmatrix}, f^K_{v_0N_1}\right) = F(\mathrm{diag}(\tilde{\omega}^{m+1},\tilde{\omega}^m,1), f_{v_0}) = F_{v_0}(m+1,m)\ .$$

We write $\beta_1(s)$ for the sum of $c_3\beta(N\gamma,s)$ over the $\gamma$ such that $N\gamma$ is quadratic over $F_{v_0}$ and for which $j = 1$ (for such $\gamma$ we have $\Delta_1(h) = 1$ unless $v_0$ is dyadic). The corresponding part of our integrand can now be put in the form

$$(2\pi i)^{-1}\int_{|t|=1} f^{\wedge}_{v_0}(s/t,t)(\beta_1(s)/t)d^{\times}t\ .$$

5.2.3. Last qudratic terms

To deal with the remaining parts of (*) and (**) denote by $\lambda$ a pair $(n_1,n_2)$ of integers and by $F_f(\lambda)$ the orbital integral $F(\mathrm{diag}(\tilde{\omega}^{n_1},\tilde{\omega}^{n_2}),f)$, where $f$ is a spherical function. If $K$ is the standard maximal compact subgroup of $GL(2,F_{v_0})$ we signify by $f_\lambda$ the characteristic function of $K\,\mathrm{diag}(\tilde{\omega}^{n_1},\tilde{\omega}^{n_2})K$, and set

$$\alpha = (1,-1) \ , \quad \langle\alpha,\lambda\rangle = n_1 - n_2 \ , \quad \lambda_j = (j,0) \qquad (j=0,1) \ .$$

Unless otherwise specified we shall assume below that $\langle\alpha,\lambda\rangle \geq 0$ . Note that $F_f(\lambda)$ and $f_\lambda$ do not depend on the order of components of $\lambda$ . The various objects here are defined with respect to $v_0$ but we delete the $v_0$ so as to simplify the notations.

Any compactly supported spherical function $f$ can be written as a finite sum of the form

$$f = \sum_{m \geq n} c_{(m,n)} f_{(m,n)} \qquad (c_{(m,n)} \text{ in } \mathbb{C}) \ ,$$

so that $f(\lambda) = c_\lambda$ . The proof of [12], Lemma 5.1, implies that for $k \geq 0$ we have

$$F_f(\lambda+k\alpha) = |\tilde{\omega}|^{-1/2\langle\alpha,\lambda+k\alpha\rangle} c_{\lambda+k\alpha} + (1-|\tilde{\omega}|) \sum_{n>0} c_{\lambda+(k+n)\alpha} |\tilde{\omega}|^{-1/2\langle\alpha,\lambda+(k+n)\alpha\rangle} \ .$$

We obtain the Plancherel formula

$$f(\lambda) = |\tilde{\omega}|^{1/2\langle\alpha,\lambda\rangle} [F_f(\lambda) - (1-|\tilde{\omega}|) \sum_{k \geq 1} a_k F_f(\lambda+k\alpha)]$$

with

$$a_1 = 1 \ , \quad a_{k+1} = 1-(1-|\tilde{\omega}|)a_k \ .$$

Put

$$b_0 = 1 \ , \quad b_k = -(1-|\tilde{\omega}|)a_k \ , \quad g = f^K_{v_0 N_1} \ .$$

It follows that

$$|\tilde{\omega}|^m \ g(m+j,m) = |\tilde{\omega}|^{m+1/2j} \sum_{k>0} b_k \ F_g(m+j+k, m-k) \ .$$

For any pair $n,r$ of integers we write $j = 0$ or $1$ if $n+r$ is even or odd. We put $m = (n+r-j)/2$ and $k = (n-r-j)/2$ so that the Satake transform

$$\hat{f}_{v_0}(s,t) = \sum_{n,r} F_{v_0}(n,r)s^n t^r$$

of $f_{v_0}$ can be put in the form

$$\sum_{m,k} \sum_{j=0,1} F_{v_0}(m+j+k,m-k)s^{m+j+k} t^{m-k} .$$

As usual we change variables in each coefficient and obtain

$$\hat{f}_{v_0}(s,t) = \sum_{m,k} \sum_{j=0,1} F_g(m+j+k,m-k)s^{m+j+k} t^{m-k}|\tilde{\omega}|^{m+(1/2)j} .$$

Hence

$$\hat{f}_{v_0}(s/t,st) = \sum_{m,k,j} F_g\ (m+j+k,m-k)\ s^{2m+j}t^{-2k-j}|\tilde{\omega}|^{m+(1/2)j}$$

and

$$(2\pi i)^{-1} \int_{|t|=1} \hat{f}_{v_0}(s/t,st)t^{2k+j}d^{\times}t = \sum_m F_g(m+j+k,m-k)s^{2m+j}|\tilde{\omega}|^{m+1/2j} ,$$

for each $j$ and $k$. It follows that for $j = 0$ or $1$ we have

$$\sum_m |\tilde{\omega}|^m\ g(m+j,m)s^m = s^{-1/2j} \sum_{k\geq 0} (b_k/2\pi i) \int_{|t|=1} \hat{f}_{v_0}(s^{1/2}/t,s^{1/2}t)t^{2k+j}d^{\times}t .$$

We can now write $\beta_{2+j}(s^{1/2})$ for the sum of

$$-c_{2+j}|\tilde{\omega}|^j\Delta_1(h)\beta(\gamma,s)s^{-1/2j}$$

over the subset of the $\gamma$ in our original integrand such that $N\gamma$ is quadratic over $F_{v_0}$ and $|h_1h_2|_{v_0} = |\tilde{\omega}|^j$. The corresponding parts of the integrand (for $j = 0,1$) are

$$\beta_{2+j}(s^{1/2}) \sum_{k\geq 0} \frac{b_k}{2\pi i} \int_{|t|=1} \hat{f}_{v_0}(s^{1/2}/t,s^{1/2}t)t^{2k+j}d^{\times}t .$$

Although the sum here contains only a finite number (depending on $f_{v_0}$) of non-zero integrals we cannot change summation and integration since $\Sigma\, b_k t^k$ does not converge on $|t| = 1$ . However for $|t| < 1$ we have

$$\sum_{k=1}^{m} a_k t^k = a_1 t + t \sum_{k=1}^{m-1} (1-a_k(1-|\tilde{\omega}|))t^k = t + t^2 \frac{1-t^{m-1}}{1-t} - t(1-|\tilde{\omega}|) \sum_{k=1}^{m-1} a_k t^k ,$$

so that

$$\sum_{k=1}^{m-1} a_k t^k = \frac{(t(1-t^m)/(1-t) - a_m t^m)}{1+(1-|\tilde{\omega}|)t} \longrightarrow t(1-t)^{-1}(1+(1-|\tilde{\omega}|)t)^{-1} .$$

Finally we consider the $\gamma$ such that $N\gamma$ lies in $A_1(F)$ modulo $NA_1(E)$ . Since $NA_1(E_{v_0})$ is $A_1(F_{v_0})$ we may assume that $F^{\wedge}(N\gamma, s, \phi_{v_0})$ is $f^K_{v_0 N_1}(N\gamma)$ . Writing $\beta_4(s)$ for the sum of $\beta(N\gamma, s)$ over all such $\gamma$ we obtain, as in the above calculations in the case $j = 0$ , that the corresponding part of the integrand is equal to

$$\beta_4(s) \sum_{k \geq 0} (b_k/2\pi i) \int_{|t|=1} f^{\wedge}_{v_0}(s^{1/2}/t, s^{1/2}t)t^{2k} d^{\times}t .$$

Since the original integral over $s$ could have been taken over two copies of the unit circle (and then divided by 2), we obtain the term indexed by $j = 0$ in the following lemma.

Note that the second summand in the displayed expressions of (5) in Lemmas 2.7, 3.11, can be dealt with as in the last comment, and therefore will not be mentioned again. To sum up the discussion we state:

LEMMA 3. <u>The difference of 5.2.1 is equal to</u>

$$(2\pi i)^{-2} \iint_{|s|=1=|t|} \beta(s,t) \hat{f}_{v_0}(s,t)\, d^\times s\, d^\times t \; +$$

$$+ \; (2\pi i)^{-2} \int_{|s|=1} d^\times s \sum_{j=0,1} \beta_j(s) \sum_{k>0} b_k \int_{|t|=1} \hat{f}_{v_0}(s/t, st) t^{2k+j} d^\times t \; .$$

<u>Here</u> $\beta(s,t)$ <u>and</u> $\beta_j(s)$ <u>are bounded integrable functions on</u> $|s| = |t| = 1$ .

5.3.1. <u>An integral expression</u>

Our efforts in the first three sections of this chapter culminate with:

LEMMA 4. <u>The right side</u> $\ell\Sigma I_0(\phi) - \Sigma I_0(f)$ <u>of Lemma 4.2 is equal to the expression displayed in Lemma 3, with new functions</u> $\beta$ <u>which are again bounded and integrable on</u> $|s| = |t| = 1$ .

The key result of this chapter, Proposition 5 below, will follow on comparing this integral expression with the (discrete) sum of Lemma 4.3, and deducing that both integral and sum are 0 . Most of the work towards the proof of the lemma was done in the previous two sections, and here we collect the remaining odds and ends. These are the remaining terms from Lemmas 2.3, 2.4, 3.7, 3.8, (5) of 2.7 and 3.11 and the limits which were added in the course of the proof of Lemma 3, and have to be deleted here. These "singular" terms are likely to be related; however our aim here is only to express the difference in question as in the lemma.

Let A denote the value of the equal scalars of third displayed lines in (1) of Lemmas 2.4 and 3.8. The corresponding terms of the trace formulae, namely,

$$\ell A \sum_h \int_{N_0(\mathbb{A})} f^K(hn)\,dn \qquad (h \text{ in } NZ(E)\backslash Z(F))$$

and

$$A \iiint \phi^K(n^{-\sigma}a^{-\sigma}n^F an)\|c/a\|^{2/\ell}\, da\,dn\,dn^F ,$$

are equal after the last contribution, from the twisted formula, was multiplied by $\ell$ . Indeed, multiplying by $\lambda_{-1}^2$ each integral can be expressed as a local product. Each $\int f_v^K(hn)dn$ is the limit of $F(a,f_v)$ as $a \longrightarrow h$ , $a$ in $A_0(F_v)$ . But $F(a,f_v)$ vanishes unless $a$ lies in $NA_0(E_v)$ , hence the limit is $0$ unless $h$ lies in $NZ(E_v)$ . If $h$ lies in $NZ(E_v)$ for all $v$ then $h$ lies in $NZ(E)$ . Hence the sum over $h$ reduces to the single term $h = 1$ . The claim now follows from the fact that $F(N\gamma,f_v) = F(N\gamma,\phi_v)$ for all $v$ and $\gamma$ in $A_0(E_v)$ .

The next difference to be considered is between the terms described by (a) second displayed lines in (1), Lemmas 2.3 and 2.4; first summand in the displayed expression of (5), Lemma 2.7, and (b) the corresponding terms from Lemmas 3.7, 3.8, 3.11, multiplied by $\ell$ .

Written out, (a) becomes

$$\ell A \sum_h \int f^K(hn) \sum \log C_v D_v dn + 2\,\ell A \phi_1(f)^{K\cap M_1}_{N_0\cap M_1}(h)$$

$$= \ell\, A \sum_h \sum_v \Big[\int f_v^K(hn)\log C_vD_v dn + 2\Phi_1(f_v)^{K\cap M_1}_{N_0\cap M_1}(h)\Big]\prod_{w\neq v} f^K_{wN_0}(h) ,$$

and (b) is

$$A \iiint \phi^K(n^{-\sigma}a^{-\sigma}\gamma n^F an\|c/a\|^{1/\ell} \sum_v \log(C_vD_v) + 2\ell\Phi_1(\phi)^{K\cap M_1}_{N_0\cap M_1}(N\gamma)$$

$$= A \sum_\gamma \sum_v \Big[\iiint \phi_v^K(\dots)\|c/a\|^{1/\ell}\log C_vD_v + 2\ell\Phi_1(\phi_v)^{K\cap M_1}_{N_0\cap M_1}(N\gamma)\Big]\prod_{2\neq v} \phi^K_{wN_0}(N\gamma) ,$$

where

$$A = 3/2(\lambda_0/\lambda_{-1} - \sum L_v'(1)/L_v(1)) .$$

The sums over $v$ are taken over a fixed finite set independent of $\phi_{v_0}$ and $f_{v_0}$.

The factor $f^K_{wN_0}(h)$ is the limit of $F(a,f_w)$ ($a$ in $A_0(F_w)$), as $a \to h$. Since $F(a,f_w)$ vanishes unless $a$ lies in $NA_0(E_w)$, $h$ lies in $NA_0(E_w)$ for all $w$ but $v$, hence $h$ lies in $NA_0(E)$ (cf. Lemma 1.1). It follows that the sum in (a) over the $h$ in $NZ(E)\backslash A_1(F)$ has to be taken only over $NZ(E)\backslash NA_1(E)$. This group is isomorphic (by the norm map) to the group $Z(E)A_1(E)^{1-\sigma}\backslash A_1(E)$, over which the sum over $\gamma$ in (b) is taken.

When considering the difference (a)-(b) we may assume that the sums over $v$ in both (a) and (b) are taken only over $v \neq v_0$. Indeed, since $\phi_{v_0}$ (and $f_{v_0}$) is spherical the functions $\Phi_1(f_{v_0})$ and $\Phi_1(\phi_{v_0})$ vanish, and the proof of Lemma 1.15 shows that when $h = N\gamma$ we have

$$\ell \int f^K_{v_0}(hn)\log C_v D_v dn = \iiint \phi^K_{v_0}(n^{-\sigma}a^{-\sigma}\gamma n^F an)\|c/a\|^{1/\ell}\log C_v D_v .$$

Note that $C_v, D_v$ on the right side are defined with respect to $E$, in contrast to those on the left which are defined with respect to $F$.

The summation formula will be applied to the function described by the sum over $v$ in (a) and the pair $NZ(E)\backslash NA_1(E)$, $NZ(\mathbb{A}_E)\backslash NA_1(\mathbb{A}_E)$, and to the function given by the sum over $v$ in (b) and the pair $Z(E)A_1(E)^{1-\sigma}\backslash A_1(E)$, $Z(\mathbb{A}_E)A_1(\mathbb{A}_E)^{1-\sigma}\backslash A_1(\mathbb{A}_E)$. That the formula can be applied follows from the properties of $\Phi_1(f_v), \Phi_1(\phi_1)$, and a study similar to that of 2.7.1/2 and 3.4.1/2 for the weighted integrals in brackets. It was already noted that the sums over $h$ and $\gamma$ in (a) and (b) include the elements of $NZ(E)$ and $Z(E)$, as the limits of our functions there are the terms from (1) of Lemmas 2.4 and 3.8. Since $f^K_{v_0N_0}(h)$ and $\phi^K_{v_0N_0}(N\gamma)$ always appears as factors in the product over $w$ in (a) and (b), arguing as in 5.1.4 we obtain that (a)-(b) is equal to

$$(2\pi i)^{-1}\int_{|s|=1} \beta(s)F^{\hat{}}(s,f_{v_0})d^{\times}s \;.$$

Here $\beta(s)$ has the usual properties and $F^{\hat{}}(s,f_{v_0})$ is equal to

$$\sum_m F_{v_0}(1,m)s^m = \frac{1}{2\pi i}\int_{|t|=1}\Big(\sum_{m,n} F_{v_0}(n,m)t^n s^m\Big)d\,t = \frac{1}{2\pi i}\int_{|t|=1} f^{\hat{}}_{v_0}(t,s)d^{\times}t \;.$$

It follows that (a)-(b) has the same form as the first displayed line in Lemma 3, but with some $\beta(s)$ replacing $\beta(s,t)$ .

To complete the proof of Lemma 4 it remains to consider the difference between (c) the sum over $h$ in $NZ(E)\backslash Z(F)$ of (2) of Lemma 2.4, and the missing limit terms of (2), Lemma 2.3, with the second summand of (5), Lemma 2.7 , and (d) the corresponding terms from Lemmas 3.7, 8, 11, multiplied by $\ell$ . Standard arguments show that the sum over $h$ has to be taken only over $h = 1$ ; the corresponding sum from (d) is already taken over $\gamma = 1$ alone. We have the usual cancellation at $v_0$ ; the component at $v_0$ is therefore always $f^K_{v_0N_1}(1)$ . We have to apply the degenerate case of the summation formula, namely the Fourier inversion formula. The Fourier transform of $f^K_{v_0N_1}$ can be calculated as in 5.2.3 by means of the Plancherel formula of that subsection. We obtain once again the term of Lemma 3 whose coefficient is $\beta_0(s)$ , only that now this coefficient is a constant. The proof of the lemma is now complete.

### 5.4.1. Traces identity

All is ready to establish the main result of this chapter.

PROPOSITION 5. In the notations of Lemma 4.2 we have

$$\ell \operatorname{tr} r(\phi') + \sum_{\eta^E} \operatorname{tr}\{I'(\sigma,\eta^E)I(\eta^E,\phi)\} = \operatorname{tr} r(f) .$$

Proof. From the equality of the expression displayed in Lemma 4.3, and the one of Lemma 3 here, we shall deduce that both are $0$ . If this is not the case we may reorder indices and assume that $\beta_0 \neq 0$ . We add to the second integral in Lemma 3 the integral

$$(2\pi i)^{-2}\int_{|s|=1} \sum_{j=0,1} \beta_j(s) \sum_{k\geq 0} b_k \int_{|t|=1} \tfrac{1}{2}[(t-1)f^{\wedge}_{v_0}(-s,-s)-(t+1)f^{\wedge}_{v_0}(s,s)]t^{2k+j}d^{\times}t d^{\times}s$$

$$= (2\pi i)^{-1}\int_{|s|=1}(-\tfrac{1}{2}\beta_0(s))[f^{\wedge}_{v_0}(s,s) + f^{\wedge}_{v_0}(-s,-s)]d^{\times}s \ , \qquad (*)$$

and subtract it later on. Since the sums over $k$ are finite we obtain

$$(2\pi i)^{-2}\int_{|s|=1} \sum_j \beta_j(s)\sum_k b_k \int_{|t|=1} [f^{\wedge}_{v_0}(s/t,st) + \tfrac{1}{2}(t-1)f^{\wedge}_{v_0}(-s,-s)$$

$$- \tfrac{1}{2}(t+1)f^{\wedge}_{v_0}(s,s)]t^{2k+j}d^{\times}t d^{\times}s \ .$$

We have

$$\sum_{k\geq 0} b_k t^{2k} = 1 - (1-|\tilde{\omega}|)t^2/(1-t^2)(1+(1-|\tilde{\omega}|)t^2) \ ,$$

so that for some $\beta'(s,t)$ our expression becomes

$$(2\pi i)^{-2}\int_{|s|=1}\int_{|t|=1} [f^{\wedge}_{v_0}(s/t,st) + \tfrac{1}{2}(t-1)f^{\wedge}_{v_0}(-s,-s) - \tfrac{1}{2}(t+1)f^{\wedge}_{v_0}(s,s)] \ \frac{\beta'(s,t)}{t^2-1} \ d^{\times}t d^{\times}s \ . \qquad (**)$$

Since $\Sigma\beta_k f^{\wedge}_{v_0}(z_k)$ of Lemma 4.3 is absolutely convergent for all $f^{\wedge}_{v_0}$, and we may take a constant $f^{\wedge}_{v_0}$, there is a positive integer $n$ such that the sum of $|\beta_k|$ over $k > n$ is less than $|\beta_0|/3$. Also we have

$$\max\{|\beta(s,t)|, |\beta'(s,t)|;\ |s|=t,\ |t|=1\} \leq c_0 ,$$

where $c_0$, as well as $c_1, c_2, \ldots$ below, denotes a fixed positive number.

Given an analytic function $\psi(z) = \Sigma a_{n_1 n_2} s^{n_1} t^{n_2}$ on the set of $z = (s,t,r)$ in $A_0(\mathbb{C})$ with $p^{-1} \leq |s|$, $|t| \leq p$, $str = \omega(\tilde{\omega})$ and $p > |\tilde{\omega}|^{-1}$, we have

$$a_{n_1 n_2} p_1^{n_1} p_2^{n_2} = \left(\frac{1}{2\pi i}\right)^2 \iint (\Sigma a_{m_1 m_2} (p_1 s)^{m_1} (p_2 t)^{m_2}) s^{-n_1} t^{-n_2}\, d^{\times}s\, d^{\times}t \qquad (|s|=|t|=1)$$

for each $n_1, n_2$, where $p_i$ is $p^{\operatorname{sgn}(n_i)}$ $(i = 1,2)$. Since $\psi$ is analytic on a compact domain it is bounded, and then

$$|a_{n_1 n_2}| \leq c_1 p^{-|n_1|-|n_2|} .$$

It follows that the finite subsums $\psi_1(z)$ defined by $|n_i| \leq N_i$ $(i=1,2)$ together with their partial derivatives of total order $\leq c_2$, uniformly approximate $\psi(z)$ and its corresponding derivatives on the subdomain $X'$ of $z$ with $|\tilde{\omega}| \leq |s|, |t|, |r| \leq |\tilde{\omega}|^{-1}$, $str = \omega(\tilde{\omega})$, and $\bar{z} \equiv z^{-1}$ (mod $W$).

The restrictions of

$$\xi(z) = \sum_w \psi(w^{-1}zw) , \quad \xi_1(z) = \sum_w \psi_1(w^{-1}zw) \qquad (w \text{ in } W) ,$$

to $X'$ are $W$-invariant and hence define functions on the space $X$ of 4.3.1. But since $\xi_1(z)$ is a finite series in $s$ and $t$ it is the Satake transform $f^{\wedge}_{v_0}$ of some $f_{v_0}$. Hence given an analytic function $\psi(z)$ as above, $\xi(z)$

together with its partial derivatives of total order $\leq c_2$ can be uniformly approximated by the power series $f^{\wedge}_{v_0}$ and its corresponding derivatives. This argument justifies the choice of the function $f^{\wedge}$ in the following paragraph.

Suppose that $\beta_0 \neq 0$ and $z_0 = (s_0,t_0)$ has a component which does not lie on the unit circle. Then for any $\varepsilon > 0$ there exists $f^{\wedge} = f^{\wedge}_{v_0}$ which is bounded by 2 on $X$, its value at $z_0$ is 1, and at $z_k$ it is bounded by $\varepsilon$ $(1 \leq k \leq n)$. Moreover we require that $f^{\wedge}(s/t,st)$ is bounded by $\varepsilon^3$ on $|s| = |t| = 1$, and that $Df^{\wedge}(s/t,st)$ is bounded by 2 on the domain described by $|s| = |t| = 1$ and $\|t-1\| < \varepsilon$ or $\|t+1\| < \varepsilon$. Here $D$ denotes the operator of partial derivation with respect to $t$. If $t$ and $t_0$ are on the unit circle then $\|t-t_0\|$ denotes their radial distance (e.g. $\|(-1)-1\| = \pi$). If $z$ and $z_0 = (s_0,t_0)$ are elements of $X$ with components on the unit circle then $\|z-z_0\|$ signifies the minimum of $\max\{\|s-s_0\|,\|t-t_0\|\}$, as $(s,t)$ runs through a set of representatives of $z$ in $X$.

Given such $f^{\wedge}$ we decompose the domain over which $t$ is taken in (**) to three subdomains, described by $\|t-1\| < \varepsilon$, $\|t+1\| < \varepsilon$, and their complement. On the first domain

$$|(f^{\wedge}(s/t,st)-f^{\wedge}(s,s))/(t-1)| \leq \max\{|Df^{\wedge}(s/t,st)|;\|t-1\|<\varepsilon\ ,|s|=|t|=1\} \leq 2\ ,$$

and $f^{\wedge}(s,s)$ (as well as $f^{\wedge}(-s,-s)$) are bounded. On the last domain $|t^2-1| \geq \frac{1}{2}\varepsilon^2$ and each of the functionsin the square brackets is bounded by $\varepsilon^3$. Similar discussion can be carried out for (*) and the first integral in Lemma 3. The conclusion is that the expression described by Lemma 3 is bounded by $c_3\varepsilon$ for $f^{\wedge}_{v_0} = f^{\wedge}$. On the other hand $|\Sigma\beta_k f^{\wedge}(z_k)|$ exceeds $\frac{1}{3}|\beta_0| - c_4\varepsilon$. For a sufficiently small $\varepsilon$ we obtain a contradiction which implies for any $k$ that if $\beta_k \neq 0$ then all components of $z_k$ lie on the unit circle.

5.4.2. End of proof

To deal with the remaining cases it suffices to consider the restrictions of the $\hat{f}_{v_0}$ to the subset of $z$ in $X$ whose components lie on the unit circle. The finite Fourier series $\hat{f}_{v_0}$ and its derivatives of order $\leq c_2$ uniformly approximate any smooth function on the subset of $z$ in $X$ with $|s| = |t| = 1$. Suppose that $\beta_0 \neq 0$ and $z_0 = (s_0, s_0)$ (with $|s_0| = 1$) in $X$. For any $\varepsilon > 0$ we can choose $\hat{f} = \hat{f}_{v_0}$ whose value at $z_0$ is $1$, which is bounded by $2$ on $X$, and whose value at $z_i$ is bounded by $\varepsilon (1 \leq i \leq n)$. We require that $\hat{f}(s/t, st)$ be bounded by $\varepsilon^3$ if $\|t-1\| \geq \varepsilon$ and $\|t+1\| > \varepsilon$, or if $t = 1$ and $\|s-s_0\| \geq \varepsilon^{3/2}$. Further we require that $D\hat{f}(s/t, st)$ is bounded by $2$ if $\|t \pm 1\| < \varepsilon$ and $\|s \pm s_0\| < \varepsilon$.

Again we decompose the domain over which $t$ is taken in (**) to three subdomains, defined by $\|t-1\| < \varepsilon$, $\|t+1\| < \varepsilon$, and their complement. On the first domain we note that $\hat{f}(s,s)$ is bounded by $2$, and that $D\hat{f}(s/t, st)$ is bounded by $2$ if $\|s-s_0\| \geq \varepsilon$ and by $2/\varepsilon^{3/2}$ if $\|s-s_0\| < \varepsilon$. Hence the contribution from this subdomain is bounded by $c_5 \varepsilon^{1/2}$. The same conclusion holds for the same reasons in the case of the second subdomain. On the third subdomain we note that the denominator of $\hat{f}(s,s)/(t-1)$ is greater than $\varepsilon$ while the numerator is bounded by $\varepsilon^3$ if $\|s-s_0\| \geq \varepsilon^{3/2}$ and by $2$ if $\|s-1\| < \varepsilon^{3/2}$. Hence this term contributes no more than $c_6 \varepsilon^{1/2}$, as does the integral of $\hat{f}(-s,-s)/(t+1)$. On $\|t+1\| \geq \varepsilon$ and $\|t-1\| \geq \varepsilon$ we have $|t^2-1| \geq \frac{1}{2}\varepsilon^2$ and $\hat{f}(s/t, st)$ is bounded by $\varepsilon^3$.

It is easy to see that (*) and the first integral in Lemma 3 are bounded by $c_7 \varepsilon$, and that $\Sigma \beta_k \hat{f}(z_k)$ is bounded below by $\frac{1}{3}|\beta_0| - c_8 \varepsilon$. We obtain a contradiction if $\varepsilon$ is sufficiently small, proving that if $\beta_k \neq 0$ then $z_k \neq (s,s)$ in $X$.

Finally we may assume that $\beta_0 \neq 0$ and $z_0$ is not of the form $(s,s)$ (modulo $W$). We may assume that $z_0 = (s_0/t_0, s_0t_0)$ for some $s_0, t_0$ on the unit circle with $t_0 \neq \pm 1$. For a sufficiently small $\varepsilon > 0$ we choose $f^\wedge$ whose value at $z_0$ is $1$, whose value at $z_i$ is bounded by $\varepsilon$ $(1 \leq i \leq n)$, which is bounded by $2$ on $|s| = |t| = 1$, such that $Df^\wedge(s/t,st)$ is bounded by $2$ if $\|t-1\| < \varepsilon$ or $\|t+1\| < \varepsilon$, and such that $f^\wedge(s/t,st)$ is bounded by $\varepsilon^3$ unless $\|(s/t,st) - z_0\| < \varepsilon^3$. We decompose $|t| = 1$ as usual and we use the last property to estimate $f^\wedge(s/t,st)$ on the domain $\|t+1\| > \varepsilon$, $\|t-1\| \geq \varepsilon$, distinguishing between the $z = (s/t,st)$ with $\|z-z_0\| \geq \varepsilon^3$ and those with $\|z-z_0\| < \varepsilon^3$. Again we obtain a contradiction which establishes that all $\beta_k$ are $0$ so that the proposition follows.

5.4.3. Reformulation

As in 4.3.2 we let $V$ be a finite set of places containing the infinite ones and those which ramify in $E$. For each $v$ outside $V$ we fix $z_v = (s_v, t_v, r_v)$ with components in $\mathbb{C}$ whose product is $\omega(\tilde{\omega}_v)$ if $v$ splits $E$ and $\omega(\tilde{\omega}^{\ell})$ if $v$ stays prime in $E$. Our applications will be derived from the following more practical form of Proposition 5.

PROPOSITION 6. For any $\phi_v$ and $f_v$ related by $\phi_v \longrightarrow f_v$ we have

$$\ell \sum \prod_{v\in V} \operatorname{tr} \pi_v^E(\phi_v') + \sum \prod_{v\in V} \operatorname{tr}\{I'(\sigma,\eta_v^E) I(\eta_v^E,\phi_v)\} = \sum \prod_{v\in V} \operatorname{tr} \pi_v(f_v) .$$

The first sum extends over all $\pi^E = \otimes \pi_v^E$ in $L_0^2(\omega_E)$ such that for each $v$ outside $V$ the component $\pi_v^E$ is unramified and $\operatorname{tr} \pi_v^E(\phi_v') = f_v^\wedge(z_v)$ for all spherical $\phi_v$. The second sum extends over the $\eta^E$ (modulo $W$) such that $\eta_v^E$ is unramified and $\operatorname{tr} I(\eta_v^E,\phi_v') = f_v^\wedge(z_v)$ (spherical $\phi_v$) for all $v$ outside

V. <u>The last sum is over the</u> $\pi = \otimes\pi_v$ <u>in</u> $L_0^2(\omega)$ <u>such that for</u> v <u>outside</u> V <u>the component</u> $\pi_v$ <u>is unramified and</u> $\operatorname{tr} \pi_v(f_v) = f_v^{\wedge}(z_v)$ <u>for all spherical</u> $f_v$ <u>obtained from spherical</u> $\phi_v$ by $\phi_v \longrightarrow f_v$ .

<u>Proof</u>. We fix v outside V and claim that

$$c(z_v) = \sum_{\pi^E} \prod_{w \neq v} \operatorname{tr} \pi_w^E(\phi_w') + \sum_{\eta^E} \prod_{w \neq v} \operatorname{tr}\{I'(\sigma,\eta_w^E)I(\eta_w^E,\phi_w)\} - \sum_{\pi} \prod_{w \neq v} \operatorname{tr} \pi_w(f_w)$$

is equal to 0 . Proposition 5 implies that

$$\sum c(z_v) f_v^{\wedge}(z_v) = 0$$

for all spherical $f_v$ obtained from spherical $\phi_v$ by $\phi_v \longrightarrow f_v$ . The sum is absolutely convergent and taken over all $z_v$ in $A_0(\mathbb{C})/W$ with determinant $\omega(\tilde{\omega}_v)$ if v splits in E and $\omega(\tilde{\omega}_v^{\ell})$ if v stays prime in E . The countable set of $z_v$ with $c(z_v) \neq 0$ is indexed by $z_i$ and we put $c_i = c(z_i)$ . We may assume that $c_0 \neq 0$ . There exists $n > 0$ so that the sum of $|c_i|$ for $i > n$ is less than $\frac{1}{4}|\beta_0|$ , and there exists $f_v$ such that $f_v^{\wedge}(z_0) = 1$ , $f_v^{\wedge}(z_i)$ is bounded by $|\beta_0|/3n$ $(1 \leq i \leq n)$, and $f_v^{\wedge}(z)$ is bounded by 2 on X . Its existence contradicts the assumption that $c(z_v)$ is non-zero for some $z_v$ . Applying induction on the set of v outside V it follows that for each finite set U disjoint from V and fixed $z_u'$ (u in U), if $c_i$ is defined as $c(z_v)$ above but with products taken (as in the proposition) only over w in V, then $(*) \sum c_i \prod f_v^{\wedge}(z_{iv}) = 0$ $(v \notin V \cup U)$. The sum is over all sequences $\{z_{iv}\}$ with $z_{iu} \equiv z_u'$ (u in U). Here we write $z \equiv z'$ if v splits (resp. stays prime) and the sets of eigenvalues of $z, z'$ (resp. $z^{\ell}, z'^{\ell}$) are distinct. Now if (e.g.) $c_0 \neq 0$, choose N with $\sum |c_i| < \frac{1}{2}|c_0|$ $(i \geq N)$, and U disjoint from V so that for each $1 \leq i < N$ there is u in U with $z_{iu} \neq z_u$. The contradiction obtained on applying (*) with this U and $z_u' = z_u$ (u in U), and $f_v^{\wedge} = 1$ for all $v \notin V \cup U$, shows that all $c_i$ are 0 , and the proposition follows.

## 6. THE CORRESPONDENCE

### 6.1.1 Lifting $\pi(\theta)$'s

In this chapter we shall prove the results about the correspondence. Both local and global results will be deduced from the equality of trace formulae in Proposition 5.6 using the local theory of Chapter 1. Various comments concerning this major equality can be made at once.

The multiplicity one and strong multiplicity one theorems for $L_0^2(\omega_E)$ (or "for G") imply that the first sum contains at most one term. By virtue of [8], Lemma 12.2, we have $\eta_1^E = w\eta_2^E$ if $\eta_{1v}^E = w_v\eta_{2v}^E$ for almost all $v$. Here $w$ (and $w_v$) denotes a rotation in $W$, and as usual we write $w\eta^E(a)$ for $\eta^E(w^{-1}aw)$. Hence the second sum contains at most one term. But $L(s,\pi^E\otimes\chi)$ is entire for all characters $\chi$ of $E^\times\backslash \mathbb{A}_E^\times$, and $L(s,I(\eta^E)\otimes\chi)$ has a pole at $s = 1$ for some $\chi$, and local L-functions do not have poles or zeros on $\operatorname{Re} s > 0$. We deduce that at most one of the two sums on the left is non-empty.

Choose $\eta^E$ modulo $W$ and a set $V$ of places such that $\eta_v^E$ is unramified for $v$ outside $V$. Let $\{z_v; v\notin V\}$ be a subset of $A_0(\mathbb{C})/W$ so that the left side becomes

$$\prod_{v\notin V} \operatorname{tr}\{I'(\sigma,\eta_v^E)I(\eta_v^E,\phi_v)\} .$$

If $\mu^E$ denotes one of the components of $\eta^E$ then we put

$$\theta = \operatorname{Ind}(W_{E/F}, W_{E/E}, \mu^E) ,$$

where $W_{./.}$ denotes the Weil group. Then $\pi = \pi(\theta)$ defines an automorphic representation, and if $\theta$ is irreducible then $\pi = \pi(\theta)$ is cuspidal in $L_0^2(\omega_E)$ (see [9], (14.2)). But for every $v$ outside $V$, and for every $\phi_v, f_v$ with $\phi_v \longrightarrow f_v$ we have

$$\operatorname{tr} \pi_v(\theta_v, f_v) = \operatorname{tr} I(\eta_v, f_v) = \operatorname{tr} I(\eta_v^E, \phi_v')$$

for some $\eta_v$ with $\eta_v(Nx) = \eta_v^E(x)$ ($x$ in $E_v^\times$). Hence $\pi(\theta)$ contributes a non-trivial term in the sum on the right of our traces equality.

LEMMA 1. <u>There is a single term on the right side and it is</u> $\pi(\theta)$.

<u>Proof</u>. Consider a $v$ outside $V$ so that $\eta_v^E$ is unramified. If $v$ stays prime in $E$ then $\eta_v^E = (\mu_v^E, \mu_v^E, \mu_v^E)$ and $\pi_v = \pi(\mu_v, \zeta_v\mu_v, \zeta_v^2\mu_v)$, where $\mu_v$ is any character of $F_v^\times$ with $\mu_v(Nx) = \mu_v^E(x)$ ($x$ in $E_v^\times$), and $\zeta$ is a non-trivial character of $F^\times N\mathbb{A}_E^\times \backslash \mathbb{A}^\times$ whose component at $v$ is denoted by $\zeta_v$. If $v$ splits into $v_1, v_2, v_3$ then $\eta_v^E = (\eta_1, \eta_2, \eta_3)$ with

$$\eta_1 = (\mu_1, \mu_2, \mu_3), \quad \eta_2 = (\mu_2, \mu_3, \mu_1) = {}^{\sigma}\eta_1, \quad \eta_3 = (\mu_3, \mu_1, \mu_2) = {}^{\sigma^2}\eta_1,$$

where $\mu_j$ $(1 \leq j \leq 3)$ are characters of $F_v^\times$, and then $\pi_v = \pi(\mu_1, \mu_2, \mu_3)$. Note that $\mu_1\mu_2\mu_3 = \chi_v$ for all $v$ outside $V$ which split in $E$, where $\chi(x) = \mu^E(Nx)$ ($x$ in $\mathbb{A}_E^\times$). Any $\pi' = \otimes\pi_v'$ enterning the sum on the right must satisfy for any $v$ outside $V$

$$\pi_v' = \begin{cases} \pi_v, & v \text{ splits in } E, \\ \pi(\zeta_v^{i_1}\mu_v, \zeta_v^{i_2}\mu_v, \zeta_v^{i_3}\mu_v), & v \text{ stays prime in } E. \end{cases}$$

The $i_j$ $(1 \leq j \leq 3)$ depend on $v$. But since there is a character $\chi'$ of $F^\times N\mathbb{A}_E^\times \backslash \mathbb{A}^\times$ with $\zeta_v^i = \chi_v'$ $(i = i_1 + i_2 + i_3)$ for all $v$ outside $V$ we deduce that $i$ is independent of $v$.

Let $\theta'$ be the representation contragredient to $\theta$, and consider the local L-functions $L(s, \pi_v' \times \pi(\theta_v'))$ and $L(s, \pi(\theta_v) \times \pi(\theta_v'))$ of [10]. They are clearly equal for $v$ outside $V$ which splits in $E$. This is also true if $v$ outside $V$ stays prime in $E$. Indeed we then see that $L(s, \pi_v' \times \pi(\theta_v'))^{-1}$

is the product over all $j$ (= 1,2,3) and $k$ (= 0,1,2) of $[1-\zeta_v(\tilde{\omega}_v)^{i_j-k}|\tilde{\omega}_v|^s]$ . But since $\zeta$ is of order 3 this product is independent of $i_j$ $(1 \leq j \leq 3)$. The theorem of [10] implies that $\pi'$ is equivalent to $\pi(\theta)$ , and the lemma follows.

We have established that for any $\phi_v, f_v$ with $\phi_v \longrightarrow f_v$ ($v$ in $V$), we have

$$\prod_{v \text{ in } V} \operatorname{tr}\{I'(\sigma,\eta_v^E)I(\eta_v^E,\phi_v)\} = \prod_{v \text{ in } V} \operatorname{tr} \pi_v(f_v) , \qquad (*)$$

where $\pi = \pi(\theta)$ .

LEMMA 2. <u>Suppose</u> $F$ <u>is a local field</u>, $E$ <u>is a cubic extension and</u> $\eta^E = (\mu^E, {}^\sigma\mu^E, {}^{\sigma^2}\mu^E)$ . <u>Then the character</u> $\chi_{I(\eta^E)}$ <u>of</u> $I(\eta^E)$ <u>exists as a locally integrable function of</u> $G \times G(E)$ <u>which is smooth on the regular set. Moreover on the regular subset of</u> $\sigma \times G(E)$ <u>it is given by</u>

$$\chi_{I(\eta^E)}((\sigma,\gamma)) = \chi_{\pi(\theta)}(h) \quad \underline{\text{where}} \quad \theta = \operatorname{Ind}(W_{E/F}, W_{E/E}, \mu^E) ,$$

<u>if</u> $h$ <u>in</u> $G(F)$ <u>is conjugate to</u> $N\gamma$ <u>and</u> $h$ <u>is regular</u>.

<u>Proof</u>. It suffices to consider the case of a unitary character $\mu^E$ . We shall deduce from (*) that

$$\operatorname{tr}\{I'(\sigma,\eta_v^E)I(\eta_v^E,\phi_v)\} = \operatorname{tr} \pi_v(f_v)$$

for all $\phi_v$ $(\longrightarrow f_v)$ when $v$ stays prime in the global field $E$ , thus establishing the lemma. The equality follows from Lemma 1.14 if $v$ splits in $E$ , and from Corollary 1.6 if $\eta_v^E$ is unramified, since then the components of $\eta_v^E$ are equal to each other. Thus we assume that $v$ stays prime in the global field $E$ . In particular $v$ is non-archimedean since $\ell = 3$ . Every character $\mu_v^E$ of $E_v^\times$ can be extended to a character $\mu^E$ of $E^\times \backslash \mathbb{A}_E^\times$ which is unramified at all (non-archimedean) places other than $v$ . Applying (*) to $\eta^E = (\mu^E, {}^\sigma\mu^E, {}^{\sigma^2}\mu^E)$ the desired equality follows since there are $\phi$ for which $\operatorname{tr}\{I'(\sigma,\eta_v^E)I(\eta_v^E,\phi_v)\}$ is non-zero.

6.2.1. Linear independence

Let $V$ be a finite set of places and suppose that for each $v$ in $V$ we have a sequence $\{\pi_{kv}\ (k \geq 0)\}$ of admissible, irreducible, infinite dimensional unitary representations of $G(F_v)$ which transform under $NE_v^\times$ by $\omega_v$.

LEMMA 3. *Suppose that for every set $\{f_v;\ v \text{ in } V\}$ of $f_v$ which transform under $NE_v^\times$ by $\omega_v^{-1}$ and which are the image of some $\phi_v$ on $G(F_v)$, the sum*

$$\sum_{k \geq 0} \prod_{v \text{ in } V} \operatorname{tr} \pi_{kv}(f_v)$$

*is absolutely convergent and its value is $0$. Then the sequence $\{\pi_{kv};v \text{ in } V\}$ is empty.*

*Proof*. Suppose that $\pi_v$ is a square-integrable representation of $G(F_v)$ which transforms under $NE_v^\times$ by $\omega_v$. By Lemma 1.2, there exists a function $f_v'$ whose orbital integrals $F_0(h,f_v')$ are $0$ for each regular $h$ in $G(F_v)$ unless $h$ is regular and it lies in $NT(E_v)$ for some elliptic torus $T$ of $G$ over $F_v$, where $F_0(h,f_v')$ is $t^{-1}\chi_{\pi_v}(h)$ with $t = |Z(F_v)\backslash T(F_v)|$. Lemma 1.9 implies that $\operatorname{tr} \pi_v'(f_v')$ is $0$ for any infinite dimensional $\pi_v'$ unless $\pi_v' \simeq \zeta_v^i \otimes \pi_v$ for some $i$. In the latter case the trace is $\ell^{-1}$ if $\pi_v \not\simeq \zeta_v^i \otimes \pi_v$ for any $i$, and $1$ if $\pi_v \simeq \zeta_v \otimes \pi_v$. Substituting $f_v'$ for $f_v$ in our sum we deduce by induction on the number of elements in $V$ that $\pi_{kv}$ is not square-integrable for any $k$ and any $v$. Hence $\operatorname{tr} \pi_{kv}(f_v)$ is not affected by the values of the orbital integrals of $f_v$ on the cubic tori.

Next we take $v$ in $V$ and fix a unitary supercuspidal representation $\tau$ of $GL(2,F_v)$ with a central character $\mu_1$. Let $\mu_2$ be a unitary character of $F_v^\times$ such that $\mu_1\mu_2 = \omega_v$ on $NE_v^\times$. Denote by $A_1^0(F_v)$ the subgroup of $a_1 = \operatorname{diag}(a,a,b)$ in $NA_1(E_v)$ with $|a|_v = |b|_v$. Let $\psi$ be a smooth function on $NA_1(E_v)$ , compactly supported modulo $NZ(E_v)$ , which transforms under $NZ(E_v)$ by $\omega_v^{-1}$ and under $A_1^0(F_v)$ by $\eta^{-1}(a_1)$ where $\eta(a_1) = \mu_1(a)\mu_2(b)$ . We choose a set of representatives $\{T\}$ for the conjugacy classes of tori in $G$ over $F_v$ . We may assume that the quadratic tori $T$ are contained in $M_1$ and the split torus is $A_0$ . For each quadratic $T$ we choose a set (denoted below by $NA_1(E_v)\backslash NT(E_v)$) of representatives for the quotient of $NT(E_v)$ by $NA_1(E_v)$ . Now there exists some $f_v = f_{v\psi}$ such that $F(h,f_v)$ vanishes for $h$ in $A_0(F_v)$ and also for $h$ in $T(F_v)$ but not in $NT(E_v)$ for any quadratic torus $T$ , and such that for every regular $h$ in $NT(E_v)$ we have

$$F(h,f_v) = \Delta(h_1)\chi_{I_{P_1}(\tau,\mu_2)}(h_1)\psi(a_1)$$

where

$$h = a_1h_1 \ , \quad h_1 \ \text{in} \ NA_1(E_v)\backslash NT(E_v) \ , \quad a_1 \ \text{in} \ NA_1(E_v) \ .$$

If $\pi_v'$ is infinite -dimensional and not square-integrable, and its central character is equal to $\omega_v$ on $NE_v^\times$ then $\operatorname{tr} \pi_v'(f_v)$ is $0$ unless $\pi_v' =$ $= I_{P_1}((\zeta^i\tau,\zeta^j\mu_2) \otimes \eta_z)$ for some integers $i$ and $j$ and a complex number $z$ . Here $\zeta$ denotes a non-trivial character of $NE_v^\times\backslash F_v^\times$ and $\eta_z$ denotes the character of $A_1(F_v)$ whose value at $\operatorname{diag}(a,a,b)$ is $|a/b|_v^z$ . Then

$$\operatorname{tr} \pi_v'(f_v) = \sum_T |W_T|^{-1} \int_{NZ(E_v)\backslash NA_1(E_v)} \psi(a)\eta(a)\eta_z(a)$$

$$\int_{NA_1(E_v)\backslash NT(E_v)} \Delta(h)^2 \bar{\chi}_{I_{P_1}(\tau,\mu_2)}(h)\chi_{\pi_v'}(h)\,dh\,da \ .$$

The sum is taken over all quadratic tori, and up to a constant ($1$ or $\ell^{-1}$) which depends only on $(\tau,\mu_2)$ we obtain

$$\psi^{\wedge}(z) = \int_{NZ(E_v)\backslash NA_1(E_v)} \psi(a)\eta(a)\eta_z(a)da .$$

Our sum can now be put in the form $\Sigma\alpha_j\psi^{\wedge}(z_j)$ $(j \geq 0)$ where $\alpha_j$ denotes the sum over $k \geq 0$ such that $\operatorname{tr} \pi_{kv}(f_v) = \psi^{\wedge}(z_j)$ of the product over $w \neq v$ in $V$ of $\operatorname{tr} \pi_{kw}(f_w)$. Since the $\pi_{kv}$ are unitary the $z_j$ lie on the imaginary axis. If $v$ is non-archimedean we may assume that $i^{-1}z_j$ lies in the interval from $0$ to $2\pi/\log|\tilde{\omega}_v|^b$ with $b = 1$ if $v$ splits or ramifies in $E$ and $b = \ell$ if $v$ stays prime in $E$, for then $\hat{\psi}(z)$ is periodic with period $2\pi i/\log|\tilde{\omega}_v|^b$.

The Fourier transforms $\psi^{\wedge}(z)$ are dense in the algebra of smooth functions (on the above domains) which, in the archimedean case, go to $0$ at infinity. Since $\Sigma\alpha_j\psi^{\wedge}(z_j)$ is absolutely convergent to $0$, if some $\alpha_j$ is non-zero (we may assume that $\alpha_0 \neq 0$ ) then there exists $n \geq 0$ such that the sum of $|\alpha_j|$ over $j > n$ is bounded by $|\alpha_0|/3$. We may choose $\psi$ such that $\psi^{\wedge}(z_0)$ is equal to $1$, $\psi^{\wedge}(z_j)$ is bounded by $\varepsilon/n$ $(1 \leq j \leq n)$ and $\psi^{\wedge}$ is bounded by $2$ over the domain of $z$. If the positive number $\varepsilon$ is sufficiently small we obtain a contradiction which implies that $\pi_{kv}$ is not of the form $I_{P_1}(\tau,\mu_2)$ with supercuspidal unitary $\tau$ for any $k$.

We can also repeat the argument with unitary $\tau$ which is the special representation $\sigma(\eta_1)$ $(\eta_1 = (\mu_1\alpha_v^{1/2},\mu_1\alpha_v^{-1/2}))$ of $GL(2,F_v)$ (resp. its complement $\pi(\eta_1)$) and deduce that for each $I_{P_1}(\zeta^i\sigma(\eta_1),\zeta^j\mu_2)$ (some $i,j$) occurring among the $\pi_{kv}$ there is a $I_{P_1}(\zeta^{i'}\pi(\eta_1),\zeta^{j'}\mu_2)$ among the $\pi_{kv}$ (resp. replace $\sigma(\eta_1)$ by $\pi(\eta_1)$ and $\pi(\eta_1)$ by $\sigma(\eta_1)$). Here $v$ can also be archimedean. Hence we may assume that the $\pi_{kv}$ are all of the form $I_{P_0}(\eta)$ (not necessarily irreducible) with a quasi-character $\eta$ of $A_0(F_v)$. It follows that $\operatorname{tr} \pi_{kv}(f_v)$

are not affected by the values of $F(h,f_v)$ on the cubic and quadratic tori.

Let $\eta^0 = (\mu_1^0,\mu_2^0,\mu_3^0)$ be a unitary character of $A(F_v) = A_0(F_v)$ such that $\mu_1^0\mu_2^0\mu_3^0 = \omega_v$ on $NE_v^\times$. Denote by $A^0(F_v)$ the subgroup of $a_0 = \mathrm{diag}(a_1,a_2,a_3)$ in $NA(E_v)$ with $|a_1|_v = |a_2|_v = |a_3|_v$. Put $\eta^0(a_0) = \prod_i \mu_i^0(a_i)$. Let $\psi$ be a smooth function on $NA(E_v)$ which is compactly supported modulo $NZ(E_v)$ and transforms under $NZ(E_v)$ by $\omega_v^{-1}$ and under $A^0(F_v)$ by $(\eta^0)^{-1}$. There is some $f_v = f_{v\psi}$ such that $F(h,f_v)$ is equal to $\Sigma\psi(w^{-1}hw)$ ($w$ in $W$) if $h$ lies in $NA(E_v)$ and to $0$ if $h$ lies in $A(F_v)$ but not in $NA(E_v)$. Put

$$\psi\hat{}(z) = \int_{NZ(E_v)\backslash NA(E_v)} \sum_{w \text{ in } W} \psi(w^{-1}hw)\eta^0(h)\eta_z(h)dh .$$

In the non-archimedean case $z = (s,t,r)$ lies in $A(\mathbb{C})/W$ and $str = 1$, and as in 4.3.1 we denote by $\eta_z$ the quasi-character of $A(F_v)$ whose value at $\mathrm{diag}(\tilde{\omega}^{n_1},\tilde{\omega}^{n_2},\tilde{\omega}^{n_3})$ is $s^{n_1}t^{n_2}r^{n_3}$. In the archimedean case $z = (s,t,r)$ lies in $\mathbb{C}^3/W$ and $s + t + r = 0$ and we denote by $\eta_z$ the character of $A(F_v)$ whose value at $\mathrm{diag}(a,b,c)$ is $|a|^s|b|^t|c|^r$.

Now if $\eta = (\mu_1,\mu_2,\mu_3)$ satisfies $\Pi\mu_i = \omega$ on $NE_v^\times$ and $\pi' = I_{P_0}(\eta)$ then $\mathrm{tr}\,\pi'(f_v) = 0$ unless $\eta = \eta^0\eta_z$ (modulo $W$) on $NA(E_v^\times)$, in which case we have $\mathrm{tr}\,\pi'(f_v) = \psi\hat{}(z)$. If $\pi$ is unitary then $z$ lies in $X$ (see 4.3.1) in the non-archimedean case. In the archimedean case for each $z$ the set of real parts of $s,t,r$ contains at least one zero and two numbers between $-1/2$ and $1/2$ whose sum is zero. Our sum can be expressed again as $\Sigma\alpha_j\psi\hat{}(z_j)$, and unless all $\alpha_j$ vanish we can choose $\psi\hat{}$ for which this absolutely convergent sum is non-zero.

It follows that $v$ can be deleted from $V$. Applying induction on the number of $v$ in $V$ the lemma follows.

Note that the proof of this lemma could have been much simpler if the space of $f_v$ which are obtained from $\phi_v$ by $\phi_v \longrightarrow f_v$ was a closed convolution algebra which contains together with each $f_v$ also $f_v^*(g) = \bar{f}(g^{-1})$ . Then a simple variant of [8], page 496 ("linear independence of characters") would have been applicable; however this is not the case.

6.2.2. Deleting places

Lemma 3 implies that if the left side of of the identity of Proposition 5.6 is empty then so is the right side. Suppose $\pi^E$ is an irreducible constituent of $L_0^2(\omega_E)$ . Then for some $V$ and a set $\{z_v\colon v \text{ outside } V\}$ the equality takes the form

$$\ell \prod_{v \text{ in } V} \operatorname{tr} \pi_v^E(\phi_v') = \sum_k \prod_{v \text{ in } V} \operatorname{tr} \pi_{kv}(f_v) . \tag{1}$$

LEMMA 4. *If for some $v$ in $V$ the representation $\pi_v^E$ is the image of some $\pi_v$ then $\pi_{kv}$ corresponds to $\pi_v^E$ for all $k$, and (1) remains valid if $v$ is deleted from $V$.*

*Proof.* The equality (1) becomes

$$\ell \zeta_1^i c \operatorname{tr} \pi_v(f_v) = \sum_k c_k \operatorname{tr} \pi_{kv}(f_v) \tag{2}$$

where

$$c = \prod_{w \in V, w \neq v} \operatorname{tr} \pi_w^E(\phi_w') , \quad c_k = \prod_{w \in V, w \neq v} \operatorname{tr} \pi_{kw}(f_w)$$

for some integer $i$ . The power of the $\ell^{th}$ root $\zeta_1$ of unity occurs since the extension of $\pi_v^E$ to $G \times G(E_v)$ may differ from the extension of $\pi_v^E$ used in the definition of the correspondence.

Let $\pi_v'$ be square-integrable and choose $f_v$ such that for any infinite dimensional $\pi_v''$ which transforms under $NE_v^\times$ by $\omega_v$ we have

$$\operatorname{tr} \pi_v''(f_v) = \begin{cases} 0, & \text{if } \pi_v'' \neq \zeta^j \otimes \pi_v' \quad \text{(all } j\text{)} \\ 1, & \text{if } \pi_v'' \simeq \zeta^j \otimes \pi_v' \text{ and } \pi_v' \simeq \zeta \otimes \pi_v' \quad \text{(some } j\text{)} \\ \ell^{-1}, & \text{if } \pi_v'' \simeq \zeta^j \otimes \pi_v' \text{ and } \pi_v' \neq \zeta \otimes \pi_v' \quad \text{(some } j\text{)} \end{cases}$$

Note that no $\pi_v$ or $\pi_{kv}$ is finite dimensional. Applying (2) with this $f_v$ we deduce from Lemma 3 that if $\pi_v$ is not of the form $\zeta^j \otimes \pi_v'$ then no $\pi_{kv}$ is of this form. Otherwise we have $\ell\, \zeta_1^i c = \Sigma c_k$, where the sum is taken over the $k$ with $\pi_{kv} \simeq \zeta^j \otimes \pi_v$ for some $j$. Hence $\Sigma\, c_k \operatorname{tr} \pi_{jv}(f_v)$ is $0$, where the sum is taken over all $k$ with $\pi_{kv} \neq \zeta^j \otimes \pi_v$ (all $j$). Lemma 3 implies that the last sum is empty.

If $\pi_v$ is not square-integrable then no $\pi_{kv}$ is. The arguments employed in the proof of Lemma 3 apply in this case as well and we deduce that each $\pi_{kv}$ corresponds to $\pi_v^E$, as required.

Now for each $v$ as in the lemma we choose the extension of $\pi_v^E$ to $G \times G(E_v)$ so that

$$\operatorname{tr} \pi_v^E(\phi_v') = \operatorname{tr} \pi_v(f_v)$$

when $\phi_v$ is the image of $f_v$. We see that $v$ can be deleted from the set $V$ of (1) as required. Note that on deleting $v$ from $V$ we add the requirement that the sums are taken over $\pi^E$ whose component at $v$ is $\pi_v^E$ and over $\pi_k$ such that $\pi_{kv}$ corresponds to $\pi_v^E$.

### 6.2.3. A single place

Note that when $V$ consists of a single element then $c = c_k = 1$, $\zeta_1^i = 1$ and the number of $\pi_{kv}$ in (1) is $\ell$.

LEMMA 5. *Suppose that the set* $V$ *of* (1) *consists of a single non-archimedean place* $v$ *which is non-split in* $E$, *and assume that* $\pi_v^E$ *is supercuspidal. Then each* $\pi_{kv}$ *is supercuspidal, there are* $\ell$ *such* $\pi_{kv}$, *and each of them corresponds to* $\pi_v^E$.

*Proof*. Since we now deal with a single $v$ it can be dropped to simplify the notations. The character $\chi_{\pi^E}$ of $\pi^E$ is known to exist as a function on the regular subset of $\sigma \times G(E)$, and we can define a function $\chi$ on the union of $NT(E)$ over a set of representatives $T$ for the conjugacy classes of tori of $G$ over $F$, by $\chi(h) = \chi_{\pi^E}((\sigma,\gamma))$ if $h$ is regular in $G(F)$ and conjugate to $N\gamma$ with $\gamma$ in $T(E)$. We have

$$\operatorname{tr} \pi^E(\phi') = \sum_T |W_T|^{-1} \int_{NZ(E)\backslash NT(E)} \chi(h)F(h,f)\Delta(h)dh ,$$

and

$$\textstyle\sum'(|W_T||Z(F)\backslash T(F)|)^{-1} \int_{NZ(E)\backslash NT(E)} |\chi(h)|^2\Delta(h)^2 dh = \ell^{-1} . \qquad (3)$$

Here and below $\sum'$ denotes a sum over a set of representatives $T$ for the conjugacy classes of elliptic tori of $G$ over $F$.

We shall now use the completeness of characters of square-integrable representations of $G(F)$. This follows from the completeness of characters of the representations of the compact group $G'(F)$, the multiplicative group of a division algebra of dimension $3^2 = 9$ over $F$, and the correspondence of representations of $G(F)$ (which can be deduced as in [4] from Corollary 2.9). The completeness statement implies that there exists a square-integrable $\pi$ so that

$$\alpha = \sum{}'(|W_T||Z(F)\backslash T(F)|)^{-1} \int_{NZ(E)\backslash NT(E)} \chi(h)\overline{\chi}_\pi(h)\Delta(h)^2 dh$$

is non-zero. Corollary 1.8 and Lemma 1.12 imply that $\pi$ is supercuspidal, and Lemma 2 implies that $\pi$ is not of the form $\pi(\theta)$.

Using his Hecke theory Jacquet has recently established that if an admisible irreducible representation $\pi$ of $G(F)$ satisfies $\pi \simeq \zeta \otimes \pi$ for a non-trivial character $\zeta$ of $NE^\times\backslash F^\times$ then ($\ell = [E:F]$ is equal to 3 and) for some quasi-character $\mu^E$ of $E^\times$ we have that $\pi = \pi(\theta)$ in the notations of Lemma 2 (private communication). Using this result we deduce from Lemma 1.5 (ii) that for our $\pi$ we have

$$\sum{}'(|W_T||Z(F)\backslash T(F)|)^{-1} \int_{NZ(E)\backslash NT(E)} |\chi_\pi(h)|^2\Delta(h)^2 dh = \ell^{-1} .$$

The Schwarz inequality implies that $|\alpha| \leq \ell^{-1}$. Substituting the function $f'$, which was constructed in the first few lines of the proof of Lemma 3, in (1), we deduce that $\ell\alpha = n/\ell$, where $n$ denotes the cardinality of the set of representations $\pi_k$ which occur on the right side of (1) and satisfy $\pi_k \simeq \zeta^i \otimes \pi$ for some integer $i$. But for each such $\pi_k$ the respresentation $\zeta \otimes \pi_k$ also appears on the right of (1) and it is inequivalent to $\pi_k$. Hence $n \geq \ell$ and $|\ell\alpha| \geq 1$ so that $\alpha = \ell^{-1}$ and $n = \ell$. We deduce from Lemma 1.9, from (3) and from the completeness of characters of square-integrable representations of $G(F)$ that $\chi = \chi_\pi$ on $NT(E)$ for all elliptic tori $T$ of $G$ over $F$.

Reordering indices we may assume that $\pi_k \simeq \zeta^k \otimes \pi$ $(0 \leq k \leq \ell)$. We shall prove that these are all of the terms on the right of (1). The proof is very similar to that of Lemma 3. As there we deduce that none of the $\pi_k$ with $k \geq \ell$ in (1) is square-integrable. In particular (1) is not affected by the values of the orbital integrals of $f$ on the cubic tori.

Next we consider a unitary representation $\tau$ of $GL(2,F)$ as in the proof of Lemma 3, and following the steps there we obtain that

$$\sum_{k\geq\ell} \operatorname{tr} \pi_k(f) = \sum_{j\geq 0} n_j \psi^{\wedge}(z_j)$$

for some set $\{z_j\}$ of distinct complex numbers; $n_j$ denotes the number of the $\pi_k$ such that $\operatorname{tr} \pi_k(f) = \psi^{\wedge}(z_j)$. In contrast to Lemma 3, we prefer here multiplicative notations and we define $\psi^{\wedge}$ for any $\psi$ as there with respect to the unramified character $\eta_z$ of $A_1(F)$ whose value at $\operatorname{diag}(\tilde{\omega}^{mb},\tilde{\omega}^{mb},\tilde{\omega}^{nb})$ is $z^{m-n}$; here again $b = \ell$ if $v$ stays prime in $E$ and $b = 1$ if $v$ ramifies in $E$. Since the $\pi_k$ are unitary the $z_j$ lie on the unit circle $|z| = 1$.

The value at $f$ of the difference between the left side of (1) and the first $\ell$ terms on the right is given by

$$\sum_T |W_T|^{-1} \int_{NZ(E)\backslash NT(E)} \Delta(h)\overline{\chi}_I(h)\psi(h)\cdot\Delta(h)(\chi - \sum_{0\leq k\leq\ell} \chi_{\pi_k})(h)dh .$$

As in the definition of $f$ we put $\psi(h) = \psi(a)$ and $(\Delta\overline{\chi}_I)(h) = (\Delta\overline{\chi}_{I_{P_1}(\tau,\mu_2)})(h_1)$ if $h = ah_1$; we write $\chi(N\gamma)$ for $\chi_{\pi^E}((\sigma,\gamma))$. The sum is taken over all quadratic tori $T$ of $G$ over $F$. Since $\pi^E$ and $\pi_k$ $(0\leq k\leq\ell)$ are supercuspidal the function $\Delta(\chi - \Sigma\chi_{\pi_k})$ is bounded on $NT(E)$ and it is compactly supported modulo $NZ(E)$. Applying the Fourier inversion formula with respect to $NZ(E)\backslash NA_1(E)$ we obtain

$$\sum_T (2\pi i|W_T|)^{-1}\int_{|z|=1}\int_t\int_a\int_h \psi(ta)(\Delta\overline{\chi}_I)(h)\Delta(ah)(\chi-\Sigma\chi_{\pi_k})(ah)\eta(t)\eta_z(t)dtdadh .$$

Here $a$ and $t$ are taken in $NZ(E)\backslash NA_1(E)$ and $h$ is in $NA_1(E)\backslash NT(E)$.

We replace $t$ by $ta^{-1}$ and define

$$\xi(z) = \sum_T |W_T|^{-1} \int_h (\Delta\bar{\chi}_T)(h) \int_a \Delta(ah)(\chi - \sum \chi_{\pi_k})(ah)\eta^{-1}(a)\eta_z^{-1}(a)\,da\,dh \ .$$

We obtain

$$(2\pi i)^{-1} \int_{|z|=1} \psi\hat{\ }(z)\xi(z)dz \ .$$

It is now easy to choose a $\psi$ for which the last integral is not equal to our absolutely convergent sum unless all $n_j$ are $0$. Hence we may assume that the $\pi_k$ $(k \geq \ell)$ in (1) are all of the form $I_{P_0}(\eta)$ (not necessarily irreducible), with a quasi-character $\eta$ of $A_0(F)$. In particular we have that $\chi = \chi_{\pi_k}$ $(0 \leq k \leq \ell)$ on $NT(E)$ for all quadratic tori $T$. Moreover we now know that (1) is not affected by the values of $F(h,f)$ on the cubic and the quadratic tori.

As in the last step in Lemma 3 we fix $\eta^0$ and for a suitable function $\psi$ as there we see that the sum of the terms indexed by $k \geq \ell$ is equal to $\Sigma\, n_j \psi\hat{\ }(z_j)$ with integral $n_j$ and with $z_j$ in $X$. Again the value at $f = f_\psi$ of the difference between the left side of (1) and the first $\ell$ terms on the right can be expressed in the form

$$(2\pi i)^{-2} \int_{|s|=|t|=1} \psi\hat{\ }(z)\xi(z)dz$$

with some bounded integrable function $\xi(z)$ on $|s|=|t|=1$. Choosing a suitable $\psi$ we deduce a contradiction in the usual way unless all $n_j$ are $0$. Hence the right side of (1) reduces to a sum over the $\pi_k$ with $0 \leq k \leq \ell$ and we see that $\chi = \chi_{\pi_k}$ on $NT(E)$ for all tori $T$, as required.

6.3.1. Existence lemma

LEMMA 6. Let $F$ be a global field, $v$ a non-archimedean place and $\tilde{\pi}_v$ a supercuspidal representation which transforms under the centre by $\omega_v$. Then there exists a cuspidal representation $\pi = \otimes\pi_w$ in $L_0^2(\omega)$ with $\pi_v \simeq \tilde{\pi}_v$ such that for each non-archimedean $w \neq v$ the component $\pi_w$ is either the Steinberg representation (for a finite number of $w$) or unramified.

Proof. Let $G'$ be a division algebra of dimension 9 (an anisotropic inner form of $G = GL(3)$) which splits at $v$ and, necessarily, at all archimedean places. Consider $f' = \otimes f'_w$ on $G'(\mathbb{A})$ whose components are smooth, compactly supported modulo the centre and transform under the centre by $\omega_w^{-1}$, such that (i) the orbital integrals of $f'_v$ are equal to those of the matrix coefficient $g \longrightarrow (\tilde{\pi}_v(g)u,\tilde{u})$ of $\tilde{\pi}_v$ (and the vectors $u,\tilde{u}$ in the space of $\tilde{\pi}_v$ and its contragredient satisfy $(u,\tilde{u})=1$), (ii) $f'_w$ is a matrix coefficient of a one-dimensional representation for all $w$ where $G'$ ramifies, (iii) $f'_w$ is the spherical function $f_w^0$ which vanishes outside $Z(F_w)K(F_w)$ at all other non-archimedean $w \neq v$, (iv) at the archimedean places $w$, $f'_w$ has a small compact support to be specified below.

The kernel of the operator $r(\bar{f}')$, where $r$ is the right representation on the space $L^2(\omega)$ of square-integrable automorphic forms on $G'(\mathbb{A})$ is given by $\sum \bar{f}'(g^{-1}\gamma h)$ ($\gamma$ in $Z(F)\backslash G'(F)$), and its trace is obtained on integrating the kernel over the diagonal $h = g$ in $Z(\mathbb{A})G'(F)\backslash G'(\mathbb{A})$. This last space is compact, and $\bar{f}'$ is compactly supported modulo the centre. Hence $\gamma$ makes a contribution to the sum of $\bar{f}'(g^{-1}\gamma g)$ only if $\gamma$ lies in a compact (modulo the centre). But $\gamma$ lies in the discrete group $Z(F)\backslash G'(F)$ too, hence in a finite set. We can now make the compact (modulo the centre) support of $\bar{f}'_w$ ($w$ archimedean) so small that only $\gamma = 1$ yields a non-zero contribution to the sum. The trace of

$r(\bar{f}')$ is therefore equal to the product of $\bar{f}'(1)$ and the volume of $Z(\mathbb{A})G'(F)\backslash G'(\mathbb{A})$ ; in particular it is non-zero.

Let $f = \otimes f_w$ be a function on $G(\mathbb{A})$ such that (a) $f_w = f'_w$ for all $w$ where $G'$ splits, (b) $f_w$ is smooth, compactly supported modulo the centre and transforms under the centre by $\omega_w^{-1}$, function on $G(F_w)$ whose orbital integrals are equal to those of a matrix coefficient of a special representation. Hence $f_w$ and $f'_w$ have equal orbital integrals for all $w$ (when a compatible choice of the elements of (ii) and (b) is made).

The choice of $f_w$ in (b) affords using Corollary 2.9, and the equality of the orbitals integrals of $f_w$ and $f'_w$ implies that the trace of the operator $r(\bar{f})$ is equal to that of $r(\bar{f}')$, which is non-zero.

On the other hand the trace of $r(\bar{f})$ is equal to the sum of $\pi'(\bar{f})$ over all irreducible constituents $\pi' = \otimes\pi'_w$ of $r$. Now (i) implies that $\operatorname{tr} \pi'_v(\bar{f}_v)$ is non-zero only if $\pi'_v = \tilde{\pi}_v$ (by the orthogonality relations of Lemma 1.9), (ii) and (b) imply that $\operatorname{tr} \pi'_w(\bar{f}_w)$ is 0 unless $\pi'_w$ is the Steinberg representation of $G(F_w)$ at all $w$ where the division algebra ramifies, and (iii) implies that $\operatorname{tr} \pi'_w(\bar{f}_w) \neq 0$ only if $\pi'_w$ is unramified. It follows that there exists an automorphic representation $\pi$ of $G(\mathbb{A})$ whose component at $v$ is the given $\tilde{\pi}_v$, it is Steinberg where $G'$ ramifies and unramified at all other finite places, as required.

Note that on considering a division algebra $G'$ which does not split over $E$, the operator $r(\bar{\phi}\times\sigma)$ and its kernel $\sum\bar{\phi}(g^{-\sigma}\gamma h)$, the same argument shows the existence of an autormophic representation of $G(\mathbb{A}_E)$ which is $\sigma$-invariant, its component at $v$ is any given $\sigma$-invariant supercuspidal local representation, and at any finite $w \neq v$ it is either Steinberg or unramified.

### 6.3.2. The lifting theorems

We can now complete the local theory.

THEOREM 7. *Suppose $F$ is a local field and $E$ is a cyclic extension of degree $\ell$. Every admissible irreducible representation $\pi$ of $G(F)$ corresponds to an admissible irreducible representation $\pi^E$ of $G(E)$ with ${}^{\sigma}\pi^E \simeq \pi^E$, and every such $\pi^E$ is obtained by the correspondence.*

To make this theorem valid we extend the definition which was used up to now by letting $\pi_{P_0}(\eta)$ correspond to $\pi_{P_0}(\eta^E)$ if $\eta$ corresponds to $\eta^E$ and letting $\pi_{P_1}'(\tau,\mu)$ correspond to $\pi_{P_1}(\tau^E,\mu^E)$ if $(\tau,\mu)$ corresponds to $(\tau^E,\mu^E)$ (here $\tau$ corresponds to $\tau^E$ either in the sense used up to now or if $\tau = \pi(\eta)$ and $\tau^E = \pi(\eta^E)$ and $\eta$ corresponds to $\eta^E$). Thus infinite dimensional $\pi$ may correspond to a one-dimensional $\pi^E$.

Note that unramified $\pi$ corresponds to unramified $\pi^E$, but the opposite direction is valid only when $E/F$ is unramified. If $\pi$ is unitary (one-dimensional) then $\pi^E$ is such, but in the opposite direction only one of the $\pi$ which correspond to $\pi^E$ has this property. If $\pi^E$ is square-integrable there are $\ell$ $\pi$'s which correspond to $\pi^E$ and they are all square-integrable. A square-integrable $\pi$ corresponds to such $\pi^E$ only if $\pi$ is not of the form $\pi(\theta)$ (see Lemma 2).

It remains to prove the theorem. It suffices to consider unitary supercuspidal $\pi^E$ and unitary supercuspidal $\pi$ not of the form $\pi(\theta)$. For such $\pi^E$ we apply the comment following Lemma 6 (with $\pi_v'=\pi^E$), (1), Lemma 4, Corollary 1.7, and Lemma 5. For such $\pi$ we apply Lemma 6 (with $\pi_v' = \pi$), (1), Lemma 4, Corollary 1.7, Lemma 3, and Lemma 5. The theorem follows.

DEFINITION. *An automorphic representation* $\pi = \otimes\pi_v$ *of* $G(\mathbb{A})$ *corresponds to an automorphic representation* $\pi^E = \otimes\pi_v^E$ *of* $G(\mathbb{A})_E$ *if* $\pi_v$ *corresponds to* $\pi_v^E$ *for all* $v$.

THEOREM 8. *Every cuspidal representation* $\pi$ *corresponds to a (unique)* $\pi^E$ *which is also cuspidal unless* $E/F$ *is cubic extension and* $\pi = \pi(\theta)$, *where* $\theta = \mathrm{Ind}(W_{E/F}, W_{E/E}, \mu^E)$ *for some character* $\mu^E$ *of* $E^\times/\mathbb{A}_E^\times$. *The* $\pi(\theta)$ *corresponds to* $I_{P_0}(\mu^E, {}^\sigma\mu^E, {}^{\sigma^2}\mu^E)$. *Each cuspidal* $\pi^E$ *with* ${}^\sigma\pi^E \simeq \pi^E$ *is obtained from* ($\ell$ *inequivalent) cuspidal* $\pi$.

*Proof*. The first claim follows from (1), Theorem 7, and Lemmas 3 and 4; the second from Lemmas 1 and 2; and the third again from (1), Theorem 7, and Lemmas 3 and 4.

## REFERENCES


1. J. Arthur, A trace formula for reductive groups, I: Duke Math J. 45 (1978), 911-952, II: Compo. Math. 40 (1980), 87-121.

2. J. Arthur, On the inner product of truncated Eisenstein series, preprint.

3. J. Arthur, The trace formula in invariant form, Ann. of Math. 114 (1981), 1-74.

4. D. Flath, A comparison of the automorphic representations of GL(3) and its twisted forms, Pacific J. Math (1981), to appear.

5. Y. Flicker, The adjoint lifting from SL(2) to PGL(3), preprint, IHES (1981).

5a. R. Godement, H. Jacquet, Zeta functions of simple algebras, Springer Lecture Notes 260 (1972).

6. Harish-Chandra, Harmonic analysis on reductive p-adic groups, Springer Lecture Notes 162 (1970).

7. Harish-Chandra, Harmonic analysis on reductive p-adic groups, Proc. Symp. Pure Math. 26 (1973), 167-192.

8. H. Jacquet, R. P. Langlands, Automorphic forms on GL(2), Springer Lecture Notes 114 (1970).

9. H. Jacquet, I. Piateski-Shapiro, J. Shalika, Automorphic forms on GL(3), Ann. of Math. 109 (1979), 169-258.

10. H. Jacquest, J. Shalika, Comparison des représentations automorphes du groupe linéaire, C.R.A.S.P., t. 284 (1977), 741-4.

11. R. Kottwitz, Orbital integrals on $GL_3$, Amer. J. Math. 102 (1980), 327-385.

12. R. P. Langlands, Base change for GL(2), Annals of Math. Study 96 (1980).

13. D. Shelstad, Orbital integrals and a family of groups attached to a real reductive group, Ann. Sci. Ec. Norm. Sup., t. 12 (1979), 1-31.

14. J.-P. Serre, Corps locaux, Hermann, Paris (1968).

15. T. Shintani, On liftings of holomorphic cusp forms, Proc. Sym. Pure Math. 33 (1979), 97-110.

## Index of terminology

## Index of notations